Shock Wave and High Pressure Phenomena

Shock Wave and High Pressure Phenomena

The Springer book series on Shock Wave and High Pressure Phenomena comprises monographs and multi-author volumes containing either original material or reviews of subjects within the field. All states of matter are covered. Methods and results of theoretical and experimental research and numerical simulations are included, as are applications of these results. The books are intended for graduate-level students, research scientists, mathematicians, and engineers. Subjects of interest include properties of materials at both the continuum and microscopic levels, physics of high rate deformation and flow, chemically reacting flows and detonations, wave propagation and impact phenomena.

The following list of subject areas further delineates the purview of the series. In all cases entries in the list are to be interpreted as applying to nonlinear wave propagation and high pressure phenomena. Development of experimental methods is not identified specifically, being regarded as a normal part of research in all areas of interest.

Material Properties Equation of state including chemical and phase composition, ionization, etc. Constitutive equations for inelastic deformation Fracture and fragmentation Dielectric and magnetic properties Optical properties and radiation transport Metallurgical effects Spectroscopy

Physics of Deformation and Flow Dislocation physics, twinning, and other microscopic deformation mechanisms Shear banding Mesoscale effects in solids Turbulence in fluids Microfracture and cavitation

Explosives Detonation of condensed explosives and gases Initiation and growth of reaction Detonation wave structures Explosive materials

Wave Propagation and Impact Phenomena in SolidsShock and decompression wave propagation Shock wave structure Penetration mechanics

Gasdynamics Chemically Reacting Flows Blast waves Multiphase flow

Numerical Simulation and Mathematical Theory Mathematical methods Wave propagation codes Molecular dynamics

Applications Material modification and synthesis Military ordnance Geophysics and planetary science Medicine Aerospace and Industrial applications Protective materials and structures Mining

More information about this series at http://www.springer.com/series/1774

Yurii Meshcheryakov

Multiscale Mechanics of Shock Wave Processes

Springer

Yurii Meshcheryakov
Institute of Problems of Mechanical Engineering
St. Petersburg, Russia

ISSN 2197-9529 ISSN 2197-9537 (electronic)
Shock Wave and High Pressure Phenomena
ISBN 978-981-16-4529-7 ISBN 978-981-16-4530-3 (eBook)
https://doi.org/10.1007/978-981-16-4530-3

This Springer imprint is published by the registered company Springer Nature Singapore Pte Ltd.
The registered company address is: 152 Beach Road, #21-01/04 Gateway East, Singapore 189721, Singapore

Preface

The concept of multiscale deformation is intensively developed from the beginning of the 80s. In this direction, the physics and mechanics of solid faces the following problems: (i) deformation and fracture flow at several scales; (ii) there is a current momentum and energy exchange between scales; (iii) there is a time-dependent velocity scattering of elementary carriers of deformation at each scale. To solve the problems, the concept of mesoscale has been introduced in several scientific schools. Nevertheless, in spite of progress in understanding the *physics* of mesoscale, a remarkable progress in *mesomechanics* of deformation processes still does not achieved. Modelling of shock wave processes in uniform media deals with the averaged characteristics of deformed solid, such as mean stress and strain. In the case of dynamic deformation, the velocity distribution of carriers of deformation at different scales should also be taken into account. The goal of monograph is to show the dominant role of mesoscale velocity non-uniformity in multiscale dynamic processes. In developing the multiscale mechanics of shock wave processes, the problems to be concerned are: (i) concept of dynamic mesoparticle as elementary carrier of deformation, (ii) mesoscale mechanisms for interscale exchange of momentum and energy, (iii) kinetic criteria for transition of solid into structure-unstable state, (IV) mesoscale mechanisms for localisation of dynamic deformation, (V) application of mesoscale mechanisms and criteria for concrete processes in solids, (VI) dynamic plasticity and fracture, dynamic recrystallisation, high-velocity penetration.

In the presented monograph, the formation of mesoscale is considered from the position of collectivisation of dislocation structure. This procedure is based on the kinetic theory of 3D dislocation continuum. The theory results in solution to 3D transport equations for dislocation structure in which both the mutual long-range interactions of dislocations and interaction with the medium are taken into account. Dynamic mesoparticles are determined as shock-induced short-living single-sign dislocation groups which create the fluctuative stress–strain tensor fields. Further, the collective characteristics of dislocations have been incorporated into governing equations to describe the shock wave behaviour of material. By using the multiscale approach, the concrete task on propagation of shock wave in heterogeneous medium has been solved analytically. In this way, the concept of the kinetic criterion for transition of dynamically deformed solid into structure-unstable state has been developed.

The transition happens when rate of change of mesoparticle velocity dispersion becomes higher than the rate of change of mean particle velocity. Above criterion is grounded on principally different mechanisms of dynamic localisation. The previously developed criteria for shock-induced localisation, such as local thermal criterion or criterion based on interference of tensile shock waves, cannot be verified experimentally because the formation of localised structures (shear bends, rotations and others) cannot be registered in real time. These structures are seen only in post-shocked specimens. The basic advantage of our experiments is a direct monitoring of behaviour of elementary carriers of deformation at the mesoscale. The latter is shown to be subdivided by two sub-levels—mesoscale-1 (1-10 μm) and mesoscale-2 (50-500 μm). In our experiments, the laser beam of interferometer which monitors the free surface of target is focused up to 50-70 μm; i.e. it registers the response of single structural element of mesoscale-2 in the form of two spatio-temporal characteristics—mean velocity of single mesoscale-2 element u(t) and velocity distribution of mesoscale-1 particles on the background of the mesoscale-2 structural element. In this case, when the velocity scattering at the mesoscale-1 achieves the critical values, the motion of mesoscale-2 structural element, as a whole, begins. This motion results from transformation of chaotic motion of meso-1 structural elements into translational motion of meso-2 element. The used experimental diagnostics allows to fix localised deformation in the form of the so-called *local velocity defect*. The velocity defect is found to be the direct evidence of localised deformation. The kinetic criterion supposes the development of mesoscale localisation because of avalanche-like increasing the rate of transferring the momentum and energy from mesoscale to macroscale as result of transformation of chaotic motions of elementary carriers of deformation at the mesoscale (mesoparticles) into translational motion of medium at the macroscale.

By using the developed theoretical and experimental approaches, the following dynamic processes in solids are investigated from the position of *mesoscale kinetics*:

1. Transition of shock-deformed materials into structure-unstable state;
2. Dynamic fracture of a series of constructional materials including titanium and aluminium alloys, copper, Armco iron, beryllium, different kinds of steels and brittle materials;
3. The shock-induced dynamic recrystallisation;
4. High-velocity penetration.

Author hopes that incorporated in monograph knowledge is logically stated and closely associated with the real behaviour of solids under dynamic loading.

Author is grateful to my colleagues who provided the experimental investigations and took part in numerous discussions on topic.

St. Petersburg, Russia
September 30 2021

Yurii Meshcheryakov

Description of the Proposed Book

The results of theoretical and experimental investigations of mechanical behaviour of solids under shock loading are considered from the position of energy and momentum exchange between mesoscale and macroscale. The theory of shock-induced dynamic mesoscale formation and criterion for transition of shock-deformed solid into structure-unstable state are developed. Relationships for coupling between particle velocity variations as a quantitative measure of intensity of shock-induced particle velocity non-uniformity at the mesoscale are incorporated into shock wave processes. The relationship for energy and momentum meso-macroexchange, together with the constitutive equations for mesoscale, is introduced. The results of theoretical investigations are compared to: (i) experimental investigations on shock loading the different kinds of solids—high-strength constructional materials, brittle materials and others, and (ii) microstructural investigations of post-shocked specimen by using optical, SEM and TEM.

The presented in monograph results are obtained in the Physics of Fracture Laboratory of Institute for Problems of Mechanical Engineering of Russian Academy of Sciences. The monograph is thought to be useful for researches who deal with the theory of multiscale mechanics of solids and engineers who are testing the materials under dynamic loading and for the post-graduated students.

Contents

Part I
Multiscale Deformation Fundamentals

Introduction

The purpose of Part I is to develop approaches that clarify the physical mechanisms of the multiscale deformation of solids. To date, a clear understanding of multiscale dynamic deformation is absent. Specifically, the following aspects are unknown:

(a) how mesoscale formation occurs in the process of dynamic deformation;

(b) how the statistical features of the mesoparticles as elementary carriers of deformation (ECD) should be taken into account.

In considering the multiscale processes in a dynamically deformed solid, the first processes studied are the deformation processes at the dislocation scale. These processes define the formation and behaviour of higher scales, such as the mesoscale and macroscale. Specifically, in the case of highly non-equilibrium processes, for adequate modelling of mesoscale formation, the polarisation of dislocation structure, the inertial features of dislocations, and the interaction of dislocations with the medium must be taken into account. In this situation, any attempt to model the multiscale dynamic deformation faces the following problems: (i) how the dynamic collectivisation of the dislocations takes into account the three-dimensional features of the dislocation structure; (ii) how to account for the transition from one scale to another while taking into account the long-range character of the dislocation interactions; (iii) how to take into account the interaction of dislocations with the medium as they move; (iv) how to account for the inertial properties of dislocations; and (v) how to account for the stochastic character of non-uniform dynamic deformation processes.

In our approach, as the fundamentals for multiscale dynamic deformation, the 3D kinetic theory of continuously distributed dislocations is developed in Part I of the monograph. By using this theory, a set of concrete problems are solved, including: (i) the decay of sub-microsecond stress pulses, (ii) the formation of mesoparticles resulting from the collectivisation of dislocations, (iii) the propagation of a shock wave in a relaxing medium and (vi) the transition from a uniform to a heterogeneous dynamic deformation.

The solution to these problems is based on the 3D kinetic description of the dislocation continuum in which the dislocation densities and flows are determined as averaged values of the dislocation velocity distribution function (1.1–1.6, 1.8). The kinetic equation for the distribution function is self-consistently linked with the stress and strain fields through the well-known continuous theory of dislocations (1.4–1.7). When applied to high-velocity deformation processes in solids, this theory takes into account: (i) the inertial properties of the ECD; (ii) the dissipative character of the dislocation motion; (iii) the long-range interaction of dislocations with each other; and (iv) the collective features of dislocations.

The development of the kinetic theory of dislocations requires the following:

1. *Definition of the velocity distribution function.* The dislocations are considered to be objects, which are characterised at each point in space by the tangent direction to the dislocation line and by the Burgers vector. Separate segments of the dislocation line may have different orientations in space and different velocities.
2. *Deduction of the kinetic equation for the velocity distribution function.* The convective and collision parts of the kinetic equation must consider both the dissipative features of the medium in which the dislocations move and their mutual long-range interactions.
3. *Deduction of the moment equations from the kinetic equation.* This system must coincide with the well-known equations of the continuous dislocation theory.

Chapter 1
The Kinetic Theory of Continuously Distributed Dislocations

Abstract Chapter 1 is devoted to formulation of the 3D kinetic theory of continuously distributed dislocations which is considered to be the basis for the mesoscale formation. In this theory, the formation of mesoscale is considered to be the dynamic polarisation and collectivization of single-sign dislocations. The dislocation densities and flows are determined as averaged values of the dislocation velocity distribution function. When applied to high-velocity deformation processes in solids, this theory takes into account: (i) the inertial properties of the ECD; (ii) the dissipative character of the dislocation motion; (iii) the long-range interaction of dislocations with each other and (iv) the collective features of dislocations. The development of the kinetic theory of dislocations includes the following steps: (i) definition of the velocity distribution function; (ii) deduction of the kinetic equation for the velocity distribution function. The convective and collision parts of the kinetic equation consider both the dissipative features of the medium in which the dislocations move and their mutual long-range interactions; (iii) deduction of the moment equations by using the successive procedure of averaging the kinetic equation for the dislocation velocity distribution function. The moment equations coincide with the well-known 3D equations of the continuous dislocation theory. Being locked with the constitutive equation for dislocation interaction with each other and with the medium where they move, the moment equations allow to describe the formation of mesoscale as totality of the short-living single-sing dislocation pile-ups.

1.1 The Dislocation Velocity Distribution Function

This chapter follows the sequence discussed in the introduction to design the kinetic theory of dislocations. Considering the configurational complexity of dislocations, it is appropriate to use the tensor description of the dislocation continuum. Such a description is used in continuous dislocation theory, where the dislocations density is the second rank tensor. The first index of the tensor characterises the tangent direction of the dislocation line, and the second index is the direction of the Burgers vector. According to this definition, $f_i k(\vec{r}, \vec{v}, t)\mathrm{d}\vec{r}\mathrm{d}\vec{v}\mathrm{d}t$ is the mathematical expectation of the number of dislocation segments of type ik in the volume $\mathrm{d}r$ at the moment from

Y. Meshcheryakov, *Multiscale Mechanics of Shock Wave Processes*, Shock Wave and High Pressure Phenomena, https://doi.org/10.1007/978-981-16-4530-3_1

t to $t + \mathrm{d}t$ with velocities from v to $v + \mathrm{d}v$. The zero moment of the distribution function is obtained from the average velocity and

$$\rho_{ik}(\vec{r}, t) = \int_{-\infty}^{\infty} f_{ik}(\vec{r}, \vec{v}, t)\mathrm{d}\vec{v} \tag{1.1}$$

yields the dislocation density tensor.

The first statistical moment of the distribution function defines the so-called dislocation velocity tensor or dislocation flow tensor:

$$J_{ij}(\vec{r}, t) = e_{ikl} \int_{-\infty}^{\infty} v_l f_{kj}(\vec{r}, \vec{v}, t)\mathrm{d}\vec{v}. \tag{1.2}$$

In this definition, the first and the second statistical moments of the distribution function coincide with the dislocation density tensor and the dislocation flow tensor introduced in the continuous theory of dislocations. The restrictions on dislocation motion in continuum theory must also be correct for the dislocation velocity distribution function. Specifically, from the definition of dislocation density,

$$\rho_{ik} = -e_{ilm}\frac{\partial w_{mk}}{\partial x_l},$$

it follows that

$$\frac{\partial \rho_{ik}}{\partial x_i} = 0.$$

Accordingly, an analogous restriction must be applied to the components of the velocity distribution function:

$$\frac{\partial f_{ik}}{\partial x_i} = 0, \tag{1.3}$$

which imposes the conservation condition for the Burgers vector along the dislocation line. The average flow of ik-dislocations in the direction p can be expressed in terms of the dislocation density tensor as

$$J_{ik} = e_{ijp}u_p\rho_{jk}, \tag{1.4}$$

where u_p is the p-component of the average dislocation velocity, which can be expressed in terms of the instantaneous velocity v and the relative velocity c:

$$C_p = V_p - u_p \tag{1.5}$$

By analogy with the kinetic theory of gas and fluids, one can introduce the statistical moments of the distribution function in the form:

$$P_{\text{ln}} = e_{ijk} \int\limits_{-\infty}^{\infty} c_l c_k f_{\text{jn}} \, \mathrm{d}\vec{v}, \tag{1.6}$$

$$Q_{mjn} = e_{jkl} e_{ils} \int\limits_{-\infty}^{\infty} c_m c_k c_s f_{\text{ln}} \, \mathrm{d}\vec{v}. \tag{1.7}$$

The value $m c_k c_l f_{\text{jn}}$ characterises carrying over the l-component of the elementary momentum in the k-direction with the dislocation segments of the kind jn, which have velocities within the interval $\vec{v}$ and $\vec{v} + \mathrm{d}\vec{v}$. In this case, $\widehat{P}$ is the analogue of the stress tensor in the kinetic theory of gas, whereas the diagonal elements determine the energy of chaotic motion on the background of the flow motion of the dislocations with the average velocity $\vec{u}$. Lastly, the curl of the third statistical moment Q_{jjn} can be identified with the flow of chaotic motion of dislocations.

The common form of the kinetic equation can be written in the form:

$$\frac{D\hat{f}}{Dt} = H_c. \tag{1.8}$$

The kinetic equation for the continuously distributed dislocations must include the redistribution of the dislocations in the volume and the change in their total number.

The left-hand side of this equation represents the convective part of the kinetic equation, whereas the right-hand side is the so-called collision part:

$$\frac{D\widehat{f}}{Dt} = \frac{\partial \widehat{f}}{\partial t} + \widetilde{N}_r \times \left(\vec{v} \times \widehat{f}\right) + \widetilde{N}_v \times \left(\frac{\partial \vec{v}}{\partial t} \times \widehat{f}\right) = H_c. \tag{1.9}$$

Here, H_c is the collision term that takes into account the nucleation, annihilation and interaction of dislocations. The components of the dislocation acceleration $\frac{\partial \vec{v}}{\partial t}$ can be determined from the equation of dislocation motion [8]:

$$m\frac{\partial \vec{v}}{\partial t} = \vec{F} - B\vec{v}, \tag{1.10}$$

where m is the "effective" dislocation mass, $\overrightarrow{F} = \hat{\sigma}\,\overrightarrow{b}$ is the Peach–Koehler force due to the external action on the dislocation, and B is the dislocation damping coefficient, which considers the interaction of moving dislocations within the medium. In the one-dimensional case, the kinetic equation has the form

$$\frac{\partial \widehat{f}}{\partial \partial t} + \vec{v}\frac{\partial \widehat{f}}{\partial \vec{x}} + \left(\frac{\vec{F}}{m} - \frac{B}{m}\vec{v}\right)\frac{\partial \widehat{f}}{\partial \vec{v}} + \frac{B}{m}\widehat{f} = H_c. \tag{1.11}$$

The convective part of the kinetic equation, due to the dependence of the acceleration of the dislocations on their velocity, differs from the comparable equation in the classical kinetics of fluids and gas, where the particles only interact with each other. Both the additional terms, $\frac{B}{m}v_x$ and $\frac{B}{m}f$, show the dependence of the dislocation motion on their dissipative interactions with the medium through the damping coefficient B. The collision term includes two components, each of which takes into account different kinds of interaction inside the dislocation continuum: $H_c = H_c^1 + H_c^2$. When only the spatial redistribution of dislocations is taken into account, excluding any change in density, it is thought to be sufficient to use the first component only. When the processes of nucleation and annihilation of dislocations are also important, both terms should be taken into account.

In our theory, the first collision term of the kinetic equation H_c^1 is introduced in the Fokker–Planck form [9]:

$$H_c^1 = -\nabla_v \times (D_1 \times \hat{f}) + \frac{1}{2}\nabla_v\nabla_v : (D_2 \times \hat{f}). \tag{1.12}$$

Here, D_1 and D_2 are the Fokker–Planck diffusion coefficients. D_1 is the dynamic friction coefficient, and D_2 is the coefficient of diffusion in the velocity space.

Now, we can write the expression for the equilibrium form of the distribution function, using the Fokker–Planck form of the collision integral. Combining (1.10) and (1.11) yields

$$\frac{\partial^2}{\partial v^2}(D_2 f) - 2\frac{\partial f}{\partial v}(D_1 f)2\left(\frac{F}{m} - \frac{B}{m}v\right) + \frac{2B}{m}f = 0. \tag{1.13}$$

In this analysis, we assume that, in equilibrium, the diffusion coefficient does not depend on the dislocation velocity, which makes it is possible to write the equilibrium equation in the form:

$$\frac{\partial^2 f}{\partial v^2} - \frac{2}{D_2}\frac{\partial f}{\partial v}\left(\frac{F}{m} - \frac{B}{m}v\right)\frac{\partial f}{\partial v} + \frac{2B}{D_2 m}f = 0. \tag{1.14}$$

If the integration constant is determined from the condition of constancy for dislocation density, the solution to Eq. (1.14) is

$$f_0(x, v) = \left(\frac{B}{D_2 m}\right)^{1/2}\rho(x)\exp\left[-\frac{B}{D_2 m}\left(v - \frac{B}{m}\right)\right]. \tag{1.15}$$

The above equation characterises the equilibrium velocity distribution function for the one-dimensional motion of dislocations. The equilibrium distribution corresponds to a mean dislocation velocity $u = \frac{F}{m} = \frac{\sigma b}{m}$ and a velocity dispersion $\langle \Delta v \Delta v \rangle = \frac{D_2 m}{B}$.

1.2 The Diffusion Coefficients of the Fokker–Planck Equation

The coefficient $D_1 = \frac{\langle \Delta v \rangle}{\Delta t}$ is called the dynamic friction coefficient. The value $m\frac{\langle \Delta v \rangle}{\Delta t}$ is a friction force in the opposite direction to the mean dislocation velocity u. In the case of the Fokker–Planck equation, the collision integral H_c includes only the mutual interactions of dislocations. The breaking force due to this interaction is of a fluctuative nature. The breaking force due to the interaction of dislocations with the medium is taken into account in the convective part of the kinetic equation.

It has been shown by Hubbard [9] that the coupling between the first and second diffusion coefficients of the Fokker–Planck equation has the form:

$$D_1 = \frac{1}{2}\frac{\partial}{\partial v}\left[\frac{\langle \Delta v \Delta v \rangle}{\Delta t}\right] = \frac{1}{2}\frac{\partial}{\partial v}(D_2). \tag{1.16}$$

Then, the collision integral can be written in an explicit form:

$$H_c^1 = \frac{\partial^2}{\partial v^2}\left(D_2 \hat{f}\right). \tag{1.17}$$

Now, our goal is to obtain an expression for D_2 with the help of the stress correlation function, $\langle \sigma\sigma \rangle$, according to the following:

$$D_2 = \frac{\langle \Delta v \Delta v \rangle}{\Delta t} = \frac{b^2}{m^2}\int\limits_{-\infty}^{\infty}\left\langle \hat{\sigma}(0,0)\hat{\sigma}(v\tau,\tau)\right\rangle \mathrm{d}t, \tag{1.18}$$

Derivation of the stress correlation function is based on the use of the continuous dislocation theory developed by Mura [6], Kosevich and Natzik [7]:

$$\rho_0\frac{\partial^2 U_m}{\partial t^2} = \frac{\partial \sigma_{mn}}{\partial x_{n;}}; \quad \frac{\partial w_{mn}}{\partial t} = \frac{\partial v_m}{\partial x} + J_{mn};$$

$$\sigma_{ik} = \lambda_{iklm} w_{lm}; \;\; \rho_{ik} = e_{ikl}\frac{\partial w_{lm}}{\partial x_m}. \tag{1.19}$$

Here, the w_{lm} are the components of the distortion tensor, U is the displacement, ρ_0 is the mass density of the medium, σ_{ik} are the components of the stress tensor, and λ_{iklm} are the elastic modulus components. The Fourier components of the displacement and stress tensors are defined by the integrals:

$$U_k(r,t) = \int\int U_k(\vec{k},\omega)e^{i\vec{k}\vec{r}-i\omega t}\,d\vec{k}d\omega$$

$$\sigma_{pq}(r,t) = \int\int \sigma_{pq}(\vec{k},\omega)e^{i\vec{k}\vec{r}-i\omega t}\,d\vec{k}d\omega.$$

In the most common case, the coupling between stress and distortion can be written in the form:

$$\sigma_{pq}(\vec{r},t) = \int_{-\infty}^{t} dt' \int_{-\infty}^{\infty} dr'\lambda_{pqmn}(\vec{r}-\vec{r}',t-t')w_{mn}(\vec{r}',t'),$$

or in Fourier notation:

$$\sigma_{pq}(\vec{k},\omega) = \lambda_{pqmn}(\vec{k},\omega)w_{mn}(\vec{k},\omega).$$

In turn, the Fourier components of the elastic distortion tensor can be written using the components of the dislocation flow tensor:

$$w_{mn}(\vec{k},\omega) = \frac{1}{\omega}(\lambda_{ijkl}G_{km}k_l k_n - \delta_{mi}\delta_{nj})J_{ji}, \tag{1.20}$$

where G_{km} is the Green's function of the dynamic theory of elasticity, and J_{ji} is the flow dislocation tensor. Accordingly, the Fourier components of stress tensor equal

$$\sigma_{pq}(\vec{k},\omega) = \frac{1}{\omega}\lambda_{pqmn}(\lambda_{ijkl}G_{km}k_l k_n - \delta_{mi}\delta_{nj})J_{ji}. \tag{1.21}$$

This allows to write the expression for the stress correlation function $\langle\hat{\sigma}(\vec{r},t)\hat{\sigma}(\vec{r}+\vec{s},t+\tau)\rangle$, which should be understood as the mean of the product of two stress values at the points $\vec{r}$ and $\vec{r}+\vec{s}$ at the moments of t and $t+\tau$. In Fourier notation, the stress correlation function equals

$$\begin{aligned}&\langle\hat{\sigma}(\vec{r},t)\hat{\sigma}(\vec{r}+\vec{s},t+\tau)\rangle\\ &= \int\int\int\int \langle\sigma(\vec{k}',\omega')\sigma(\vec{k},\omega)\rangle \exp\left[i\vec{k}'\vec{r}+i\vec{k}(\vec{r}+\vec{s})\right.\\ &\left.-i\omega' t - i\omega(\tau+t)\right]d\vec{k}'d\omega'd\vec{k}d\omega.\end{aligned}$$

In Fourier notation, this integral can be written using δ-functions:

$$\left\langle \hat{\sigma}(\vec{k}', \omega')\sigma(\vec{k}, \omega)\right\rangle = (\hat{\sigma}\hat{\sigma})_{\vec{k}\omega}\delta(\vec{k}+\vec{k}')\delta(\omega+\omega'), \qquad (1.22)$$

from which

$$\langle\sigma(\vec{r}, t)\sigma(\vec{r}+\vec{s}, t+\tau)\rangle = \int (\hat{\sigma}\hat{\sigma})_{\vec{k}\omega} e^{i\vec{k}\vec{r}-\omega t}\, \mathrm{d}\vec{k}\mathrm{d}\omega. \qquad (1.23)$$

Expression (1.23) can be considered the definition of the value $(\vec{\sigma}\vec{\sigma})_{\vec{k}\omega}$, which is the Fourier amplitude of the correlation function. When $s = 0$ and $\tau = 0$, the value $(\vec{\sigma}\vec{\sigma})_{\vec{k}\omega}$ is the spectral density of the mean square of the stress fluctuations:

$$\left\langle\hat{\sigma}\hat{\sigma}\right\rangle = \int\int \left(\hat{\sigma}\hat{\sigma}\right)_{\vec{k}\omega} \mathrm{d}\vec{k}\mathrm{d}\omega. \qquad (1.24)$$

The theory of dynamic elasticity allows a link between the stress correlation function and the correlation function of the dislocation flow:

$$\left\langle \sigma(\vec{k}', \omega')\sigma(\vec{k}, \omega)\right\rangle = -\frac{\eta\eta}{\omega^2}\left\langle \hat{J}(\vec{k}', \omega')\hat{J}(\vec{k}, \omega)\right\rangle, \qquad (1.25)$$

where

$$\eta_{pqij} = \lambda_{pqmn}(\lambda_{ijkl}k_l k_m - \delta_{mj}\delta_{nj}). \qquad (1.26)$$

Introducing the spectral density of the mean-square density for the flow dislocation tensor, by analogy with (1.24), one obtains

$$\left(\hat{\sigma}\hat{\sigma}\right)_{\vec{k}\omega} = -\frac{\hat{\eta}\hat{\eta}}{\omega^2}(\hat{J}\hat{J})_{\vec{k}\omega}. \qquad (1.27)$$

The value $\left(\widehat{J}\widehat{J}\right)_{\vec{k}\omega}$ can be found using the dislocation velocity distribution function in the following manner. The components of the flow dislocation tensor at the point $\vec{r}$ at the moment t are determined as follows:

$$J_{pq}(\vec{r}, t) = e_{pmn}\sum_j \rho_{mq} v_n^j \delta(\vec{r} - \vec{r}_j(t)),$$

where r is the position of the j-dislocation at the moment t, and v_n^j is the velocity of the j-dislocation. In Fourier notation, this expression can be written in the form

$$J(\vec{k}, \omega) = \frac{b^2}{(2\pi)^4} j \sum \vec{v}^j \int\int e^{-i\vec{k}\vec{r}_j(t')+i\omega t}\, \mathrm{d}t\mathrm{d}\vec{r}.$$

Then, the Fourier amplitude of the flow dislocation tensor is

$$\langle JJ\rangle = \frac{b^4}{(2\pi)^8}\sum_j \vec{v}^j\vec{v}^j \int\int\int\int \Big\langle \exp\Big[-i\vec{k}'\vec{r}_j(t') - i\vec{k}'\vec{r}_j(t) + i\omega t' + i\omega' t\Big]\Big\rangle \mathrm{d}t\mathrm{d}t'\mathrm{d}\vec{r}\mathrm{d}\vec{r}'$$

In this equation, the terms related to different kinds of dislocation can be excluded. Let us express the displacements from a position $\vec{r}_j(t)$ to a position $\vec{r}_j(t')$ through dislocation velocity $\vec{v}_d$. Taking into account that identical kinds of dislocations cannot be distinguished, one obtains

$$(JJ) = \frac{b_\theta^2}{(2\pi)^8}\int\int\int\int \vec{v}^2 f_\theta(\vec{v})\Big\langle \exp\Big[-i(\vec{k}+\vec{k}')\vec{r}(t) + i(\omega+\omega')t + i(\omega - \vec{k}\vec{v})\tau\Big]\Big\rangle \mathrm{d}\vec{v}\mathrm{d}t\mathrm{d}\tau\mathrm{d}\vec{r}$$

where $\tau = t - t'$.

Replace the exponential terms in the integral with δ-functions. Then, the correlation function for dislocation flow tensor has $\Big\langle J(\vec{k}',\omega')J(\vec{k},\omega)\Big\rangle = \frac{b_\theta^2}{(2\pi)^3}\delta(\vec{k}+\vec{k}')\delta(\omega+\omega')\int v^2 f_\theta(\vec{v})\delta(\omega - \vec{k}\vec{v})\mathrm{d}\vec{v}$ the form:

Under the conditions $\vec{k} = \vec{k}'$ and $\omega = \omega'$, one obtains the spectral density of the mean-square fluctuations for the dislocation flow tensor:

$$\big\langle (JJ)_{\vec{k}\omega}\big\rangle = \frac{b_\theta^2}{(2\pi)^3}\int f_\theta(\vec{v})v^2\delta(\omega - \vec{k}\vec{v})\mathrm{d}\vec{v}. \tag{1.28}$$

Then, the spectral density of the mean-square fluctuations for stress equals

$$\big(\hat{\sigma}\hat{\sigma}\big)_{\vec{k}\omega} = -\frac{b^2}{\omega^2}(\hat{\eta}\hat{\eta})\int v^2 f_\theta(\vec{v})\delta(\omega - \vec{k}\vec{v})\mathrm{d}\vec{v}. \tag{1.29}$$

Lastly, after joining Eqs. (1.21), (1.23) and (1.29), the diffusion coefficient is

$$D_2 = -\frac{b^4}{(2\pi)^3}\frac{\Delta t}{m^2}\int\int \frac{(\eta\eta)}{k^2} f_\theta(\vec{v})\,\mathrm{d}\vec{v}\mathrm{d}\vec{k}. \tag{1.30}$$

Consider the simplest case, when the motion of the dislocations occurs in a single plane, and the tensor $\big(\hat{\eta}\hat{\eta}\big)$ can be simplified to μ^2. Then, (1.30) is reduced to

$$D_2 = -\frac{b^3}{(2\pi)^3}\frac{\Delta t}{m^2}\rho\int_{-\infty}^{\infty}\frac{\mathrm{d}\overleftarrow{k}}{k^2}, \tag{1.31}$$

where ρ is the mean dislocation density according to the definition. To avoid non-physical divergence in the core of the dislocation, the cutting procedure should be carried out on the wave vector. In this case, the wave vector k_0 may be taken at the integration limit. This wave vector is determined by the relationship of $k_0 b \gg 1$. Then, the diffusion coefficient becomes

$$D_2 = \frac{b^4}{(2\pi)^3} \frac{\Delta t}{m^2} \rho, \tag{1.32}$$

The interval Δt can be found from the equation for the motion of dislocations (1.10). The solution to this equation yields the mean dislocation velocity:

$$u_d = \frac{\sigma b}{B}(1 - e^{-\frac{t}{\Delta t}}), \tag{1.33}$$

where

$$\Delta t = \frac{m}{B} \tag{1.34}$$

Then, the diffusion coefficient takes the form

$$D_2 = \frac{1}{(2\pi)^3} \frac{b^4}{\Delta t B^2} \mu^2 \rho. \tag{1.35}$$

The dispersion of the dislocation velocity equals

$$D_{2v} = \langle \Delta v_d \Delta v_d \rangle = \frac{1}{(2\pi)^3} \frac{b^4}{\Delta t B^2} \mu^2 \rho, \tag{1.36}$$

which coincides with the expression obtained by Alekseev and Strunin [10].

1.3 Transport Equations

When solving specific problems in dynamic plasticity, a detailed description of the elementary processes at the microscale that takes the velocity distribution into account is not always necessary. There are many dynamic processes in solids that can be solved at the so-called hydrodynamic scale. The macroscopic equations deal with the averaged characteristics of the inner structure in the form of the so-called representative volume. If instead, the moment equations for ECD are obtained by averaging the kinetic equation for the ECD, one obtains the solution at the hydrodynamic scale for the ECD only. In this case, to link the behaviour of the ECD with the macroscopic values, such as mean stress and strain, the continual theory of dislocations must be used. At the beginning, the mean densities and mean flow velocities for the ECD are found, and after that, the macroscopic variables are deduced. The transport equation for dislocations can be obtained from the kinetic equation using the formal procedure commonly used in the mechanics of gas and fluids. In accordance with this procedure, the moment equations are obtained by multiplying the kinetic equation by the reciprocal degree of particle velocity and integration in velocity space.

To obtain the zero statistical moment of the kinetic equation, the latter must be multiplied by the zeroth degree of the dislocation velocity, that is, by unit, and integrated over the velocities. During the formal procedures of integration, it is necessary to consider the tensor character of the dislocation velocity distribution function and remember the impossibility of moving along the dislocation line. The integration of the left (convective) part of the kinetic equation yields

$$\begin{aligned} &\text{(a)} \int \tfrac{\partial \hat{f}}{\partial t} \mathrm{d}\vec{v} = \tfrac{\partial}{\partial t} \int \hat{f} \mathrm{d}\vec{v} = \tfrac{\partial}{\partial t} \hat{\rho}; \\ &\text{(b)} \int \tfrac{\partial}{\partial \vec{v}} \times \left(\vec{v} \times \hat{f}\right) \mathrm{d}\vec{v} = 0; \\ &\text{(c)} \int \tfrac{\partial}{\partial \vec{r}} \times \left(\vec{v} \times \hat{f}\right) \mathrm{d}\vec{v} = \tfrac{\partial}{\partial \vec{r}} \times \int \left(\vec{v} \times \hat{f}\right) \mathrm{d}\vec{v} = \tfrac{\partial}{\partial \vec{r}} \times \hat{J}. \end{aligned} \tag{1.37}$$

The second term disappears as the distribution function equals zero at the integration limits. For the collision term,

$$\int \frac{\partial}{\partial \vec{r}} \times \left(\vec{v} \times \hat{f}\right) \mathrm{d}\vec{v} = \frac{\partial}{\partial \vec{r}} \times \int \left(\vec{v} \times \hat{f}\right) \mathrm{d}\vec{v} = \frac{\partial}{\partial \vec{r}} \times \hat{J}\, H_c^2.$$

one obtains

$$\begin{aligned} &\text{(a)} \int \alpha \left(\vec{v} \times \hat{f}\right) \mathrm{d}\vec{v} = \alpha \hat{J}; \\ &\text{(b)} \int \beta \hat{f}(\vec{v}') \mathrm{d}\vec{v}' \int \hat{f}(\vec{v}) \mathrm{d}\vec{v} = \beta \hat{\rho} \hat{\rho}, \end{aligned} \tag{1.38}$$

so the collision term has the form

$$H_c^2 = \alpha \hat{J} - \beta \hat{\rho} \hat{\rho}. \tag{1.39}$$

To obtain the Fokker–Planck part of the collision term H_c^1, let us multiply the expression (1.11) by the function $\varphi(\vec{v})$ and integrate it in the velocity space:

$$\begin{aligned} &\int \varphi(\vec{v}) \left[-\frac{\partial}{\partial v_\mu} (D_{1\mu} f_{ij}) + \frac{1}{2} \frac{\partial}{\partial v_\mu} \frac{\partial}{\partial v_\nu} (D_{2\nu\mu} f_{ij}) \right] \mathrm{d}\vec{v} \\ &= \int \frac{\partial \varphi}{\partial v_\mu} (D_{1\mu} f_{ij}) \mathrm{d}\vec{v} - \frac{1}{2} \int \frac{\partial \varphi}{\partial v_\mu} \frac{\partial}{\partial v_\nu} (D_{2\nu\mu} f_{ij}) \mathrm{d}\vec{v} \end{aligned}$$

The terms in the right-hand side of this equation are equal to zero because of the relationship (1.16) between the diffusion coefficients. This means that the long-range interaction between dislocations in the Fokker–Planck collision term does not influence macroscopic transport of the momentum and energy. This transport is described by the second part of the collision term H'. Summarising Eqs. (1.37) and (1.38) yields the ultimate form of the zero moment for the kinetic equation:

$$\frac{\partial \widehat{\rho}}{\partial t} + \frac{\partial}{\partial v} \times \widehat{J} = \alpha \widehat{J} - \beta \widehat{\rho} \widehat{\rho} \tag{1.40}$$

Equation (1.40) coincides with the analogous equation in continuum dislocation theory. If the sources and stops of dislocations are absent, this equation corresponds to the Burgers vector conservation law.

In order to derive the transport equation corresponding to the second statistical moment of the kinetic equation, one must multiply the latter by $\vec{v}$ and integrate in velocity space. The first term on the left-hand side gives

$$\int \vec{v} \times \frac{\partial \hat{f}}{\partial t} \mathrm{d}\vec{v} = \frac{\partial}{\partial t} \int \left(\vec{v} \times \hat{f}\right) \mathrm{d}\vec{v} - \int \left(\vec{\dot{v}} \times \hat{f}\right) \mathrm{d}\vec{v}. \tag{1.41}$$

To determine the acceleration of the dislocations, $\vec{\dot{v}}$, in the second term, one can use the equation of dislocation motion (1.10) to obtain

$$\int \vec{v} \times \frac{\partial \hat{f}}{\partial t} \mathrm{d}\vec{v} = \frac{\partial}{\partial t} \hat{J} - \frac{1}{m} \hat{\sigma} \times \hat{\rho} + \frac{B}{m} \hat{J}. \tag{1.42}$$

$$\int \vec{v} \times \left(\frac{\partial}{\partial \vec{r}} \times (\vec{v} \times \hat{f}) \right) \mathrm{d}\vec{v} = -\frac{\partial}{\partial \vec{r}} \hat{P} - \frac{\partial}{\partial \vec{r}} \left[\vec{u} (\vec{u} \times \hat{\rho}) \right], \tag{1.43}$$

where $\widehat{P}$ is the kinetic stress tensor, and $\vec{u}$ is the mean dislocation velocity. The third term in the left-hand side of the equation is as follows:

$$\int \vec{v} \times \left[\frac{\partial}{\partial \vec{v}} \times (\vec{v} \times \hat{f}) \right] \mathrm{d}\vec{v} = \frac{2}{m} \hat{\sigma} \times \hat{\rho} - \frac{2}{m} B \hat{J}. \tag{1.44}$$

The collision term of the equation after multiplying by $\vec{v}$ and integration in velocity is

$$\alpha \int \vec{v} \times (\vec{v} \times \hat{f}) \mathrm{d}\vec{v} = \alpha (\hat{S} + \vec{u} \times \hat{J}). \tag{1.45}$$

$$\beta \int \vec{v} \times \hat{f}(\vec{v}) \times \int \hat{f}(\vec{v}') \mathrm{d}\vec{v} \mathrm{d}\vec{v}' = \beta \hat{J} \hat{\rho}, \tag{1.46}$$

where

$$\hat{S} = \int \vec{c} \times (\vec{c} \times \hat{f}) \mathrm{d}\vec{v} \tag{1.47}$$

and $\vec{c} = \vec{v} - \vec{u}$ is the relative velocity of the dislocations. Summarising the right and left sides, one obtains for the first moment equation:

$$\frac{\partial \widehat{J}}{\partial t} = \frac{\partial}{\partial \vec{r}} \cdot \left(\widehat{P} + \vec{u}\widehat{J}\right) + \frac{1}{m}\left(\hat{\sigma} \times \hat{\rho} - B\widehat{J}\right) = \alpha\left(\widehat{S} + \vec{u} \times \widehat{J}\right) - \beta\widehat{J} \times \hat{\rho}. \tag{1.48}$$

In the absence of sources and stops of dislocations, the transport equation for the momentum is as follows:

$$\frac{\partial \vec{u}}{\partial t} \times \hat{\rho} - \frac{\partial}{\partial \vec{r}} \times \widehat{P} - \frac{\partial}{\partial \vec{r}}(\vec{u} \times (\vec{u} \times \hat{\rho})) + \frac{1}{m}\left[(\hat{\sigma} \times \hat{\rho}) - B\vec{u} \times \hat{\rho}\right] = 0. \tag{1.49}$$

This equation is the analogue of the well-known equation for momentum transport in a two-phase medium. In our case, one phase is the dislocation structure, and the other phase is the crystalline lattice.

Let us deduce the equation of the second moment of the kinetic equation. For that, we must multiply twice by the velocity vector and integrate over the velocities. As a result, one obtains

$$\begin{aligned} &\left(\frac{\partial}{\partial t} - \frac{B}{m}\right)\left[\widehat{S} - \vec{u} \times (\vec{u} \times \hat{\rho})\right] + \frac{2}{m}[\vec{u} \times (\hat{\sigma} \times \hat{\rho})] - \frac{1}{m}\hat{\sigma} \times (\vec{u} \times \hat{\rho}) \\ &\quad + \frac{\partial}{\partial \vec{r}}\left\{\widehat{Q} + \vec{u}\left[\widehat{S} + \vec{u} \times (\vec{u} \times \hat{\rho})\right] + \widehat{P}_1\right\} \\ &\quad = \alpha\left\{\vec{u} \times \left[\widehat{S} + \vec{u} \times (\vec{u} \times \hat{\rho})\right] + \widehat{T} - 2\vec{u} \times \widehat{P}\right\} - \beta\left[\widehat{S} + \vec{u} \times (\vec{u} \times \hat{\rho})\right] \times \hat{\rho}. \end{aligned} \tag{1.50}$$

Here, the definitions for $\widehat{Q}$ and $\widehat{S}$ are used.

If $\widehat{S}$ is the energy of chaotic motion of dislocations in random stress fields, then the sum $\left[\widehat{S} + \vec{u} \times (\vec{u} \times \hat{\rho})\right]$ can be related to the common energy transported by dislocations. The second term multiplied by the dislocation mass in a sense directs the transport of the kinetic dislocation energy. As the value $\widehat{Q}$ characterises the transport of the chaotic energy (the analogue of heat flow vector in the kinetic theory of gas), Eq. (1.50) reflects the energy balance in the system of moving dislocations. The interaction with the crystalline lattice is described by the term $\frac{B}{m}\left[\widehat{S} + \vec{u} \times (\vec{u} \times \hat{\rho})\right]$. The action of external forces on the dislocation is reflected by the term

$$\frac{2}{m}[\vec{u} \times (\hat{\sigma} \times \hat{\rho})] - \frac{1}{m}\hat{\sigma} \times (\vec{u} \times \hat{\rho}).$$

The above transport equations obtained for dislocations in a crystalline lattice describe the deformation process at the scale level of ECD. When known, the characteristics of the dislocation motion can be used to determine the macroscopic variables through the continuum theory of dislocation. However, the equation system (1.40)–(1.50) is non-locked until an additional relationship between the kinetic tensor $\widehat{P}$ and the dislocation density (an analogue of the equation of state in the mechanics of fluids and gases) is present. As a locking equation, the relationship describing the long-range interactions of dislocations can be used:

$$P = -D^2 \oint_{x_1} \frac{\rho(s,t)}{x-s} \mathrm{d}s \tag{1.51}$$

In our approach, the emphasis is on two aspects: (i) the type of rheological coupling between the dislocation and the medium and (ii) the kind of relationship between dislocations.

Let us write the one-dimensional version of energy transport in the case of the absence of sources and stops of dislocations:

$$\left(\frac{\partial}{\partial t} - \frac{B}{m}\right)(S + u^2\rho) + \frac{1}{m}\sigma\, b\, u\, \rho + \frac{\partial}{\partial x}\left[Q + u\left(S + u^2\rho\right) + P_1\right] = 0 \tag{1.52}$$

or

$$\frac{\partial}{\partial t}\left(S + u^2\rho\right) + \frac{1}{m}\sigma\, b\, \rho\, u - \frac{B}{m}\left(S + u^2\rho\right) + \frac{\partial}{\partial x}\left[Q + u(S + u^2\rho) + P_1\right] = 0. \tag{1.53}$$

The term $\left(S + u^2\rho\right)$ is the total energy of dislocation motion. Here, S is the energy of the chaotic motion of the dislocation, and $u^2\rho$ is the energy of the translational motion of the dislocations with the mean velocity u.

According to Eq. (1.53), the energy balance includes the following parts: (i) the expenditure of energy of external forces, (ii) the ductile breaking in the medium where the dislocations have moved, (iii) the energy of the interaction of the dislocations with each other and (iv) heat transportation. In the presence of sources and stops of dislocations, the right-hand side of the balance equation characterises the expenditure of the external forces on activation of sources and stops.

References

1. Fertziger J, Cooper G (1976) Mathematical theory of transport processes in gases. 534
2. Frenkel YaI (1948) Statistical physics. Edition of Academy of Acience of USSA, p 760
3. Zorski H (1968) Statistical theory of dislocations. Int J Solids Struct 4:959–974
4. Kosevich AM (1978) Dislocations in elasticity theory. Nauka, p 256

5. Krener E (1968) Initial study of plasticity theory based on statistical mechanics. In: Colloquium on inelastic behavior of solis. Battelle Institute, Columbus, Ohio, September 1968
6. Mura T (1963) Continuous distribution of moving dislocations. Phil Mag 8:843–853
7. Kosevich AM, Natzic VD (1966) Breaking of dislocation in a medium with dispersion of elastic modules. Fizika Tverdogo Tela 4:1250–1259
8. Kosevich AM (1962) Equation of dislocation motion. J Exp Theor Phys 42:637–648 (Russian edition)
9. Hubburd J (1960) The friction and diffusion coefficients of the Fokker-Plank equation. Proc Roy Soc London A 260:114–126
10. Alekseev AA, Strunin BM (1975) Viscous breaking of dislocations in random stress fields. In: "Dislocation dynamics" Kiev. Naukova Dumka, pp 132–137

Chapter 2
Decay of Sub-microsecond Stress Pulses

Abstract The system of two nonlinear integral equations for the mean dislocation density and dislocation flow is solved for the case when relaxation time for dislocation structure is inversely proportional to the mean density and velocity of dislocations. The spatio-density profile is shown to become steeper if the amplitude of load stress and dislocation velocity dispersion are increased. To solution to the problem of decay of sub-microsecond stress pulses, the relaxation form of the kinetic equation for the continuously distributed dislocation is used. Solution to integral equation under condition $w/k \gg v_0$ results in expression for the decay decrement which takes into account for the dislocation velocity distribution. The expression for decrement is used for the determination of drag coefficient of dislocations. The decrease of dynamic viscosity during sub-microsecond pulse durations correlates with the results of experimental investigations on the mobility of edge dislocations in NaCl and LiF. It has been shown that the coefficient of dynamic viscosity decreases fivefold during the transition to nanosecond impact velocities.

2.1 Introduction

In this chapter, the kinetics of moving dislocations is applied to shock wave propagation. We consider two problems: (i) the influence of the mesoparticle velocity distribution on the spatio-temporal profiles of shock waves and (ii) the influence of the mesoparticle velocity distribution on the decay of stress pulses in the medium. Both problems are very important in the development of new materials, especially nanomaterials. On these scales, the fluctuative stress fields initiated by the velocity distribution of the dislocations can cause a resonance interaction with the collective oscillations of the dislocations, which, in turn, can lead to unexpected catastrophic phenomena in real situations. In addition, the second problem concerns the interaction of debris with the screens of cosmic apparatus with very short stress pulse duration.

Y. Meshcheryakov, *Multiscale Mechanics of Shock Wave Processes*, Shock Wave and High Pressure Phenomena, https://doi.org/10.1007/978-981-16-4530-3_2

2.2 Dislocation Kinetics and Structure of Shock Waves

In an attempt to link the dislocation velocity distribution with the structure of shock waves, there are two restrictions: (i) the velocity equilibrium is established much sooner than the spatial equilibrium [1], and (ii) the collision term in the right-hand side of the kinetic equation for the dislocation velocity distribution function is in the relaxation form [2, 3]. This non-equilibrium kinetic equation is as follows:

$$v\frac{\partial f}{\partial x} = \frac{f - f_0}{\tau}, \tag{2.1}$$

where τ is the structural relaxation time of the dislocation ensemble. Equation (2.1) must be included in the expressions for the first and second statistical moments of the velocity distribution function:

$$n(x) = \int\limits_{-\infty}^{\infty} f(x, v)\mathrm{d}v, \tag{2.2}$$

$$n(x)v(x) = \int\limits_{-\infty}^{\infty} vf(x, v)\mathrm{d}v. \tag{2.3}$$

The formal solution to Eq. (2.1) can be written in the form

$$f(x, v) = \frac{1}{\tau v}\int\limits_{0}^{x} f_0(\xi, v)\exp\left(\frac{\xi - x}{\tau v}\right)\mathrm{d}\xi. \tag{2.4}$$

The equilibrium velocity distribution function for moving dislocations has been obtained in Chap. 1 in the form

$$f_0(x, v) = \left(\frac{B}{D_2 m}\right)^{1/2} \rho(x)\exp\left[-\frac{B}{D_2 m}\left(v - \frac{B}{m}\right)\right]. \tag{2.5}$$

Here, D_2 is the velocity dispersion, B is the drag coefficient of the dislocation interactions with the medium, and m is the mass of the dislocation. This distribution function takes into account the inertial features of dislocations, the velocity distribution and their interaction with the medium in which they move. Taking into account Eqs. (2.2), (2.3) and (2.5), one obtains the following equation system of two integral equations:

$$n(x) = \sqrt{\frac{2\beta}{\pi} \int\limits_{-\infty}^{\infty} \int\limits_{-\infty}^{x} \frac{n(x)}{\tau v} \exp\left[\frac{(\xi - x)}{\tau v} - \beta(v - u)^2\right] \mathrm{d}\xi \, \mathrm{d}v}, \tag{2.6}$$

$$n(x)v(x) = \sqrt{\frac{2\beta}{\pi} \int\limits_{-\infty}^{\infty} \int\limits_{-\infty}^{x} \frac{n(x)}{\tau} \exp\left[\frac{(\xi - x)}{\tau v} - \beta(v - u)^2\right] \mathrm{d}\xi \, \mathrm{d}v}, \tag{2.7}$$

where

$$\beta = \frac{B}{D_2 u}. \tag{2.8}$$

This obtained system is nonlinear, contains non-locality and seems to be too complex to solve analytically. However, in the case of a deformed solid, the dependence of the structural relaxation time on the dislocation velocity and density can be used to simplify the system. Under strain rates higher than $10^3 \mathrm{c}^{-1}$, there is a linear dependence of flow stress on the strain rate [4]:

$$\sigma = \sigma_0 + \eta \dot{\gamma}, \tag{2.9}$$

where η is the viscosity linked with the time of relaxation as $\eta = \tau \mu$ (μ is the shear modulus). From the positions determined from the dislocation dynamics, the plastic strain rate is determined by Orowan's equation as follows:

$$\dot{\gamma} = bnu, \tag{2.10}$$

where n and u are the mean density and mean velocity of the moving dislocations, respectively, and b is the value of the Burgers vector. Thus, the structural relaxation time is inversely proportional to the mean density and the mean velocity of dislocations as follows:

$$\tau = \frac{(\sigma - \sigma_0)}{\mu bnu}. \tag{2.11}$$

Taking Eq. (2.11) into account, Eqs. (2.6) and (2.7) acquire the form

$$n(x) = \chi \int\limits_{-\infty}^{\infty} \int\limits_{-\infty}^{x} \frac{n^2(\xi)u(\xi)}{\tau v} \exp\left[\frac{\chi(x - \xi)\eta(\xi)u(\xi)}{v} - \beta(v - u)^2\right] \mathrm{d}\xi \, \mathrm{d}v, \tag{2.12}$$

$$n(x)u(x) = \chi \int_{-\infty}^{\infty} \int_{-\infty}^{x} n^2(x)u(\xi) \exp\left[\frac{\chi(x-\xi)n(\xi)u(\xi)}{v} - \beta(v-u)^2\right] d\xi dv \tag{2.13}$$

Decomposing the exponent into the Taylor set on the variable $\lambda = \frac{v}{u}$ in the vicinity of $\lambda = 1$ yields

$$\chi(\xi - x)n(\xi) - \chi(\xi - x)n(\xi)(\lambda - 1) + \left[\chi(\xi - x)n(\xi) - \beta u^2(\xi)\right](\lambda - 1)^2 \tag{2.14}$$

The first term of this decomposition corresponds to the so-called hydrodynamic approximation when all the dislocations move with identical velocities. In this case, the system (2.12)–(2.13) transforms into two independent equations as follows:

$$n(x) = \chi \int_0^x n(\xi) \exp[\chi(\xi - x)n(\xi)] dx, \tag{2.15}$$

$$n(x)u(x) = \chi \int_0^x n^2(\xi)u(\xi) \exp[\chi(\xi - x)n(\xi)] dx. \tag{2.16}$$

Solving Eq. (2.16) using successive approximations leads to the following expression for the mean density of dislocations in a shock wave as follows:

$$n(x) = \chi\left[\exp\left(-\frac{kx^2}{4}\right)\right]\left[\frac{\sqrt{\pi k^2}}{8}x^2 + \frac{kx}{2} + \sqrt{\frac{\pi k}{4}}\right] \tag{2.17}$$

Thus, taking the dependence of relaxation time on density and velocity of dislocations into account, even in hydrodynamic approximation, results in a velocity profile shape of $n(x) : \exp(-x^2)$. In order to account for the dislocation velocity distribution, in Eq. (2.15), the square terms must be ignored. Omitting the intermediate procedure, one obtains a final expression for the dislocation density in the form $n(x) \cong \exp\left(-\frac{k\beta x^2}{\chi}\right)$. Therefore, the density profile becomes steeper if the amplitude of the load stress and the velocity dispersion of the dislocations are increased. The analogous form has the mean velocity profile. It can be shown that an exponential shape for the profile corresponds to an independent relaxation time for the velocity and density of dislocations.

2.3 Decay of Sub-microsecond Stress Pulses

The qualitative description of dynamic stress relaxation from the position given by dislocation dynamics was first established by Smith [5]. It has been shown that the shock wave front contains the charged dislocation surface. In the absence of that surface, the crystalline lattice is subjected to extremely high compression stresses (see Fig. 2.1a), which exceed the theoretical strength of the crystal. But if the moving dislocation surface is behind the shock front, stress relaxation occurs (Fig. 2.1b).

It was later shown by Weertman [6] that the longitudinal and transverse waves radiate into the region behind the shock front. The sources of these waves are the dislocations at the Smith's surfaces. These waves create the rapidly fluctuating stress fields that interact with the dislocation structure of the crystal. Thus, the following qualitative picture of stress relaxation can be imagined: it is generated in front of the high densities of charged dislocations, which create the quickly fluctuating stress fields from interactions with the collective oscillations of the dislocation structure, which result in the decay of stress pulses. The velocity distribution of the dislocations is thought to be essential to understand this kind of decay.

For the quantitative description of the relaxation process, let us use the equations of the continuum theory of dislocations which link the mean densities and mean velocities of the dislocations with the stresses and displacements in a medium as follows [7]:

$$\rho \frac{\partial v_k}{\partial t} = \frac{\partial \sigma_{ik}}{\partial x}, \tag{2.18}$$

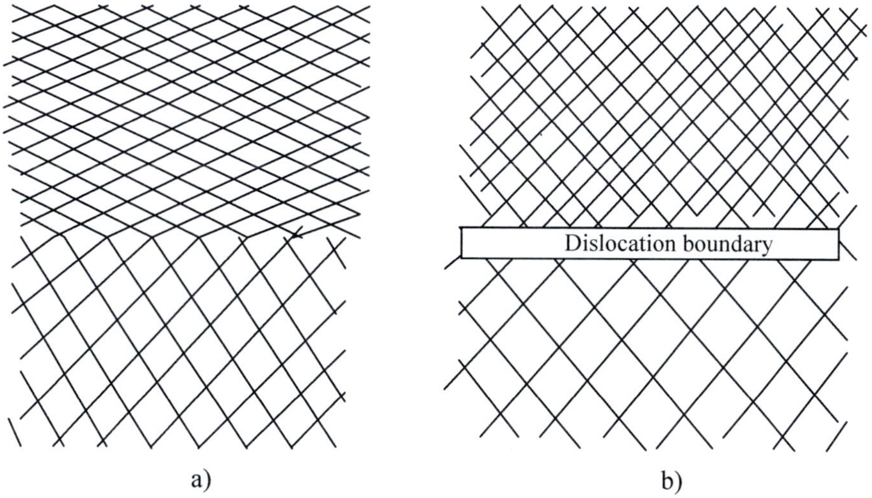

Fig. 2.1 Propagation of a shock front through a crystal: **a** without stress relaxation, **b** with stress relaxation according to Smith's model

$$\frac{\partial v_k}{\partial x_i} = \frac{\partial w_{ik}}{\partial t} - J_{ik}, \tag{2.19}$$

$$\sigma_{ik} = \lambda_{iklm} w_{lm}. \tag{2.20}$$

where v_k are the components of the particle velocity, w are components of the distortion tensor, J_{ik} are the components of the flow tensor of the dislocations, and ρ is the density of the medium. Equations (2.19)–(2.21) can be united into the second-order differential equation as follows:

$$\frac{\partial^2 \sigma_{ik}}{\partial x_i \partial x_l} - \rho \lambda_{iklm} \frac{\partial \sigma_{im}}{\partial t^2} + \rho \sum_{\Theta} \frac{\partial J_{ik}^{(\Theta)}}{\partial t} = 0. \tag{2.21}$$

Here Σ indicates summation over all the kinds of dislocations. In Fourier notation, all variables are of the form exp(ikx-ωt). Then, Eq. (2.22) transforms into

$$k_i k_l \sigma_{ik} - \rho \omega^2 \lambda_{iklm} \sigma_{lm} - i \sum_{\Theta} J_{ik}^{(\Theta)} = 0. \tag{2.22}$$

The tensor dislocation flow can be written through the dislocation velocity distribution function (see Chap. 1) as follows:

$$J_{ik}^{(\Theta)} = \varepsilon_{mli} \tau_l b_k \int_{-\infty}^{\infty} v_m^{(\Theta)} f_{lk}^{(\Theta)}(x, v, t) \mathrm{d}v. \tag{2.23}$$

Here $f_{lk}^{(\Theta)}$ are the components of the dislocation velocity distributed function introduced in Chap. 1. For the determination of the distribution function, the relaxation form of the kinetic equation can be used as follows [2, 3]:

$$\frac{\partial f^{(\Theta)}}{\partial t} + v_i \frac{\partial f^{(\Theta)}}{\partial x_i} + \varepsilon_{ikl} \tau_l b_k m_{mn}^{-1} \frac{\partial f^{(\Theta)}}{\partial v_m} = \frac{f_0 - f^{(\Theta)}}{\tau} \tag{2.24}$$

where m is the effective mass of dislocation, τ is the relaxation time for the dislocation velocity distribution function, and f_0 is the equilibrium distribution function. Let us consider the distortion component corresponding to the plane of sliding of a line of dislocation and σ_{ik} as the component of shear stress. In this case, the components of the distribution function must correspond to the dislocations of direction τ_i and Burgers vector b_k. We also omit the symbols for different sliding planes (Θ), which indicates the kind of dislocation. In a quasi-equilibrium situation, the dislocation velocity distribution function can be presented in the form

$$f = f_0 + f, \tag{2.25}$$

$$f' = f_0. \tag{2.26}$$

Then, from Eq. (2.25), one obtains, in Fourier notation

$$i\omega f' + ikvf' + \frac{1}{\tau}f' = -\sigma bm^{-1}\frac{\partial f_0}{\partial v}, \tag{2.27}$$

from which

$$f' = i\sigma bm^{-1}\frac{\partial f_0}{\partial v}\left(\omega - kv - \frac{i}{\tau}\right)^{-1}. \tag{2.28}$$

For the single sliding system, the tensor of dislocation flow is determined as follows:

$$J = b\int\limits_{-\infty}^{\infty} vf'\mathrm{d}v. \tag{2.29}$$

Then, Eq. (2.23) takes the form

$$k^2 - \frac{\rho}{\mu}\omega^2 + \rho\omega\frac{b^2}{m}\int\limits_{-\infty}^{\infty}\frac{\partial f_0}{\partial v}\left(\omega - kv - \frac{i}{\tau}\right) = 0 \tag{2.30}$$

One can further use the equilibrium velocity distribution function in the form

$$f_0(x, v) = \left(\frac{B}{D_2 m}\right)^{1/2}\rho(x)\exp\left[-\frac{B}{D_2 m}\left(v - \frac{B}{m}\right)\right]. \tag{2.31}$$

Substituting Eq. (2.32) into (2.31) yields

$$1 - \frac{\omega^2}{k^2 c_t^2} + \frac{c_{\text{ph}}^2}{c_t^2}\frac{\omega_0^2}{v_0^3}\int\limits_{-\infty}^{\infty}\frac{v(v - u_d)\exp\left[-\frac{(v-u_d)^2}{v_0^2}\right]}{\omega - kv - \frac{i}{\tau}}\mathrm{d}v - \frac{c_{\text{ph}}^2}{c_t^2} = 0. \tag{2.32}$$

where the following definitions are introduced:

$c_t = \sqrt{\frac{\rho}{\mu}}$ is the velocity of the shear waves,

$c_{\text{ph}} = \frac{\omega}{k}$ is the phase velocity of the oscillations,

$\omega_0 = \sqrt{\frac{\rho b\mu}{m}}$ is the frequency of the collective oscillations of the dislocation structure in the crystal, and

$v_0 = \sqrt{\frac{mD}{2B}}$ is the mean velocity of the fluctuative motion of the dislocations.

The integral in Eq. (2.33) can be presented as the sum of two integrals, each of which is the average of the values $v^2\left(\omega - kv - \frac{i}{\tau}\right)^{-1}$ and $v\left(\omega - kv - \frac{i}{\tau}\right)^{-1}$ over the equilibrium velocity distribution (2.32)

$$1 - \frac{\omega^2}{k^2 c_t^2} + i\frac{2c_{\text{ph}}^2}{c_t^2}\frac{\omega}{k}[\langle y_2\rangle - u_d\langle y_3\rangle] - \frac{c_{\text{ph}}^2}{c_t^2} = 0, \tag{2.33}$$

where

$$\langle y_2\rangle = \frac{1}{kv_0\sqrt{\pi}}\int_{-\infty}^{R}\frac{v^2}{\left(\omega - kv - \frac{i}{\tau}\right)}\exp\left[-\frac{(v-u_d)^2}{v_0^2}\right]\mathrm{d}v, \tag{2.34}$$

$$\langle y_3\rangle = \frac{1}{kv_0\sqrt{\pi}}\int_{-\infty}^{R}\frac{v}{\left(\omega - kv - \frac{i}{\tau}\right)}\exp\left[-\frac{(v-u_d)^2}{v_0^2}\right]\mathrm{d}v. \tag{2.35}$$

After averaging, one obtains

$$\langle y_2\rangle = \frac{\omega^2}{ik^2}\left[\frac{\sqrt{\pi}}{kv_0}\right]\exp\left[-\left(\frac{\omega}{kv_0} - \frac{u_d}{v_0}\right)^2\right] + \frac{i}{\omega}\left[1 + \frac{1}{2}\frac{1}{\left(\frac{\omega}{kv} - \frac{u_d}{v_0}\right)}\right] \tag{2.36}$$

$$\langle y_3\rangle = 0. \tag{2.37}$$

Then Eq. (2.34) becomes

$$1 - \frac{\omega^2}{k^2c_t^2} + 2i\sqrt{\pi}\left(\frac{\omega}{kv_0}\right)^3\exp\left[-\left(\frac{\omega_0}{kv_0} - \frac{u_d}{v_0}\right)\right] = 0. \tag{2.38}$$

The frequency of oscillation can be presented in the form

$$\omega = \omega_0(1 - i\delta). \tag{2.39}$$

An approximation of $\frac{\omega_0}{k} \gg v_0$ yields

$$\begin{aligned}\delta &= \sqrt{\pi}\left(\frac{\omega_0}{kv_0}\right)^3\exp\left[-\left(\frac{\omega_0}{kv_0} - \frac{u}{v_0}\right)^2\right]\frac{c_{\text{ph}}}{c_t}\\ &= \sqrt{\pi}\left(\frac{c_{\text{ph}}}{c_t}\right)^3\exp\left[-\left(\frac{\omega_0}{kv_0} - \frac{u_d}{v_0}\right)^2\right]\frac{c_{\text{ph}}}{c_t}\end{aligned} \tag{2.40}$$

Consider now two limits corresponding to, first, zero viscous drag on dislocations ($B = 0$) and, second, large values for the drag coefficient ($B = \infty$). In the first case, the exponent tends to zero since the average dislocation velocity is much greater than the mean fluctuative velocity ($u_d \gg v_0$), that is,

$$\frac{u_d}{v_0} = \frac{\sigma b}{B}\left(\frac{2B}{D_2 m}\right)^{1/2} \to \infty \tag{2.41}$$

$$\frac{\omega_0}{k v_0} = \frac{\omega_0}{k}\left(\frac{2B}{D_2 m}\right)^{1/2} \to 0 \tag{2.42}$$

As a result, $\delta \to 0$, that is, decay is absent. In the second case, $\frac{u_d}{v_0} \to 0$, so the decay is also zero. When the mean fluctuative velocity increases, the decay decrement decreases, which shows that the velocity distribution of dislocation leads to decreased stress pulse decay.

In the compression wave, the dislocation velocities can be both greater and smaller than the mean velocity. In the ensemble of particles distributed according to (2.32), the number of dislocations with velocities greater than the number of dislocations with the velocities smaller the mean velocity is smaller. Therefore, the number of dislocations transported away exceeds the number of dislocations the higher velocities. As a result, the decay of waves takes place.

The above expression for shock wave decay can be used to determine the drag coefficient B. The coupling of the drag coefficient with the dislocation velocity distribution is determined by Eq. (1.35) based on the kinetic theory of continuously distributed dislocations. However, by definition, the mean fluctuative velocity of dislocations equals

$$v_0 = \sqrt{\frac{mD_2}{2B}}. \tag{2.43}$$

Comparison of Eqs. (1.35) and (2.44) yields

$$B = \sqrt{\rho}\frac{\mu b^2}{v_0} \tag{2.44}$$

The mean fluctuative velocity of dislocations can be determined from the experimental curves for stress pulse decay in [8]. The data necessary for calculations can be taken from Table 2.1.

According to the data, the frequency of the collective oscillations of dislocation structure equals $\omega_0 = 10^{10}\mathrm{c}^{-1}$. The duration of the pulse front equals $\tau_{fr} = 10^{-8}$s, which corresponds to a basic harmonic wavelength of $\lambda = 3.3 \times 10^{-4}$ cm and a wave vector of $k = 10^4$. Then the phase velocity of the collective oscillations is $c_{ph} = \frac{\omega_0}{k} = 5 \times 10^5$ cm/s. For further calculations, we can use the experimental curve seen in Fig. 2.2 from [8]. The decay of the stress pulse can be presented in the form

$$\sigma = \sigma_0 \exp[-(\omega_0 \delta)t] \tag{2.45}$$

Table 2.1 Physical characteristics of Armco iron and A-95 aluminium

Content	Armco iron	A-95 aluminium
Longitudinal sound velocity, cm/s	6×10^5	6.4×10^5
Elastic precursor equilibrium amplitude, GPa	0.68	0.053
Density, g/cm^3	7.85	2.7
Dislocation drag coefficient,dyn $\times$ s/cm^2	8×10^{-4}	5.6×10^{-4}
Burgers vector, cm	2.85×10^{-8}	2.86×10^{-8}
Shear modulus, GPa	81.4	25.64

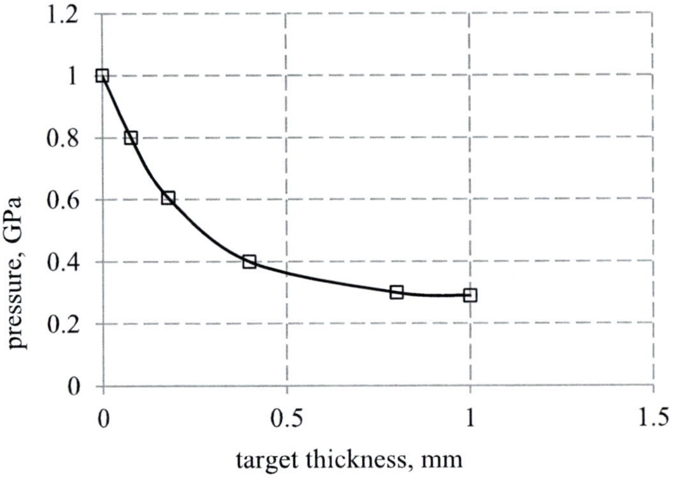

Fig. 2.2 Decay of a sub-microsecond stress pulse in A-95 aluminium

From the decay curve in Fig. 2.2, the value of the decay decrement equals $\delta = 7 \times 10^{-4}$. Then from (2.41), $\frac{c_{\text{ph}}}{v_0} = 3.6$ and the mean fluctuative velocity of dislocation equals $v_0 = 1.43 \times 10^5$ cm/s. Substitution into Eq. (2.41) yields $B = 2 \times 10^{-5}$ ps. This value is smaller by one order of magnitude than the value obtained in experiments using stress pulse durations in the region of microseconds [9]. However, the decrease of dynamic viscosity during sub-microsecond pulse durations obtained from the above estimates correlates with the results of experimental investigations on the mobility of edge dislocations in NaCl and LiF [10]. It has been shown that the coefficient of dynamic viscosity decreases fivefold during the transition to nanosecond impact velocities.

References

1. Vlasov AA (1978) Nonlocal statistical mechanics. Nauka, p 264 (Russian edition)
2. Fertziger J, Cooper G (1976) Mathematical theory of transport processes in gases, p 534
3. Kihara T, Aono O (1963) J Phys Soc Japan 18:837–851 (1963)
4. Al'shitz VI, Idenbom VL (1975) Dynamic drag of dislocations. In: "Dislocation dynamics" Kiev. Naukova Dumka. 211–217 (1975). (Russian edition)
5. Smith CS (1958) Metallographic studies of metals after explosive shock. Trans Metal Soc AIME 814:5974–6599
6. Weertman JW (1973) Dislocation mechanics in high stress rates. Metallographical effects in high stress rates. Pergamon Press, N.Y
7. Kosevich AM (1978) Dislocation in the elasticity theory. Nauka, p 256
8. Nedbai AI, Sud'enkov YuV, Rozhin GV, Filippov NM (1980) Influence of loading velocity on the behavior of elastic-plastic metals. Lett J Tech Phys 6(18):19–21
9. Gorman TA, Wood DS, Vreeland T (1969) Mobility dislocations in aluminum. J Appl Phys 40:833–843
10. Al'shitz VI, Idenbom VL (1975) Dynamic drag of dislocations. In: "Dislocation dynamics" Kiev. Naukova Dumka, pp 232–275. (Russian edition)

Chapter 3
The Collectivisation of Dislocations and Formation of Mesoscale

Abstract The mechanism of formation of the dynamic mesostructure is developed. The approach is based on solutions to the transport equations for moving dislocations which are obtained from the kinetic equation for the tensor velocity distribution function. The equations include both the interactions of dislocations with the medium and their long-range interactions with each other. The purpose of this chapter is to show that, under dynamic deformation, the formation of stable dislocation groups is possible, and they can be regarded as mesoparticles, that is, mesoscale elementary carriers of deformation. At the initial stage of dislocation collectivisation, the equation system for the dislocation density and momentum incorporates only the processes for the redistribution of dislocations. The second stage of dislocation collectivisation includes the processes of multiplication and annihilation of dislocations, as well as the redistribution of dislocations. The processes of dislocation collectivisation are solved using a simulation procedure for the transport equations for the dislocation structure. The solution concerns the step-like loading of A-95 aluminium. The simulation shows that the space profiles for dislocation density are non-monotonous. The moment $t = 45$ ns corresponds to the initiation of mesoparticle formation when the density of dislocations had not reached the maximum value. At the moment $t = 60$ ns, the relative dislocation density reaches the maximum value of $\frac{\rho}{\rho_0} = 1.4$. Then, the structure begins to "wash-away", so by $t = 150$ ns, the dislocations are uniformly distributed in the medium. The lifetime of the mesoparticle is between 150 and 200 ns.

3.1 Introduction

The transport equations obtained in Chap. 1 for the dislocation structure can be used to build a new approach to dynamic plasticity. This approach is based on solutions to the transport equations for moving dislocations. The equations include both the interactions of dislocations with the medium and their long-range interactions with each other. As the relaxation time for dislocations, $\tau \cong 10^{-8} - 10^{-9}$ s, is very small compared to macroprocesses, it is possible to consider the dislocation processes as stationary.

Y. Meshcheryakov, *Multiscale Mechanics of Shock Wave Processes*, Shock Wave and High Pressure Phenomena, https://doi.org/10.1007/978-981-16-4530-3_3

The purpose of this chapter is to show that, under dynamic deformation, the formation of stable dislocation groups is possible, and they can be regarded as mesoparticles, that is, mesoscale ECD. At the initial stage of dislocation collectivisation, the equation system for the dislocation density and momentum incorporates only the processes for the redistribution of dislocations. The second stage of dislocation collectivisation includes the processes of multiplication and annihilation of dislocations, as well as the redistribution of dislocations. The conditions for the formation of stable dislocation groups, which depend on the loading stress and material parameters, are found analytically, whereas the processes of dislocation collectivisation are solved using a simulation procedure.

3.2 Formation of Dynamic Mesostructures

The equation system for the dislocation density and momentum in the one-dimensional case is as follows:

$$\frac{\partial \rho}{\partial t} + \frac{\partial}{\partial x}(\rho u_d) = 0; \tag{3.1}$$

$$\rho \frac{\partial u_d}{\partial t} + \rho u_d \frac{\partial u_d}{\partial x} + \frac{\partial P}{\partial x} - \frac{2}{m}(\sigma b - B u_d) = 0. \tag{3.2}$$

Here, ρ is the dislocation density, u_d is the mean velocity of dislocations, σ is the external stress, b is the Burgers vector, P is the stress due to the interaction of dislocations with each other, and B is the drag coefficient. Equations (3.1) and (3.2) are obtained from the kinetic equation for the tensor velocity distribution function, so the one-dimensional case corresponds to the behaviour of single-sign dislocations. In this situation, the equation system (3.1)–(3.2) describes grouping single-sign dislocations, which results in the formation of mesostructures.

In the second equation, the stress tensor P, in agreement with the character of the interactions of dislocations in an elastic medium, can be written in the form of a Cauchy integral (Eq. (1.51). For step-like loading of the medium, the boundary conditions take the form:

$$\begin{aligned} &1)\quad u_d(x,0) = 0 \\ &2)\quad \rho(x,0) = \rho_0 \\ &3)\quad \sigma(0,t) = \sigma(t) \end{aligned} \tag{3.3}$$

In dimensionless form, the system (3.1)–(3.2) can be written as:

$$\frac{\partial r}{\partial t} + a_s \frac{\partial}{\partial x}(r u_d) = 0;$$

$$r\frac{\partial u_d}{\partial t} + a_s \times r u_d \frac{\partial u_d}{\partial x} + b_s \times r u_d = c_s f(x) + d_s \times r;$$

$$f(x) = \frac{\partial}{\partial x}\int\limits_{x_1'}^{x_2'} \frac{r(s,t)}{x-s}\mathrm{d}s. \tag{3.4}$$

Here, $a_s = \frac{u_0 t_0}{x_2}$; $b_s = \frac{2}{m}Bt_0$; $c_s = \frac{Dt_0}{x_0 u_0 \rho_0}$; $d_s = \frac{2}{m}t_0\frac{\sigma b}{u_0}$; $u_0 = \frac{\sigma b}{B}$; $v = \frac{u_d}{u_0}$.

$$\frac{\partial r_i}{\partial t} = a_i r_i + b_i;\ b_i = a_s r_{i-1}\frac{u_{i-1}}{h}.$$

where u_i is the value of the previous iteration. After decomposing the exponent into sets in the vicinity of t_j, one obtains:

In the present chapter, the modified Picard method has been used [1, 2]. This technique is often applied for the solution of hydrodynamic problems. It is a highly stable method, which allows wave profiles with the large gradients of initial data to be calculated. Here we present the first equation of the system (2.4).

$$\frac{\partial r_i}{\partial t} + a_s\frac{r_i u_i - r_{i-1} - u_{i-1}}{h} = 0;$$

A linear approximation yields:

$$\frac{\partial r_i}{\partial t} = a_i r_i + b_i,$$

where

$$r_i^{j+1} = r_i^j + \sum_{s=1}^{n} a_i^{s-1}\frac{\Delta t_j^s}{s!}\left(a_i r_i^j + b_i\right).$$

By analogy with the second equation of (3.4):

$$u_i^{j+1} = u_i^j \sum_{s=1}^{n} c_i^{s-1}\frac{\Delta t_j^s}{s!}\left(c_i u_i^j + d_i\right).$$

The following Cauchy integral is used:

$$\int\limits_{x_1}^{x_2}\frac{r(s)}{x-s}\mathrm{d}s = \ln\left|\frac{x_2 - x}{x - x_1}\right| + \sum_{k=1}^{m}\frac{r(x_k) - r(x)}{x_k - x}\Delta x \tag{3.5}$$

This solution concerns the step-like loading of A-95 aluminium. The damping coefficient B is used as the fitting parameter. The typical space profiles of dislocation density for different values of damping coefficient are provided in Figs. 3.1 and 3.2. The figures allow the kinetics of mesoparticle formation due to the collectivisation of single-sign dislocations to be traced. The simulation shows that the space profiles for dislocation density are non-monotonous. The moment $t = 45$ ns corresponds to the initiation of mesoparticle formation when the density of dislocations had not reached the maximum value. At the moment $t = 60$ ns, the relative dislocation density reaches

Fig. 3.1 Space profiles for $B = 10^{-5}$ $n\textit{з}$ at the times: 1—45 ns, 2—60 ns, 3—90 ns, 4—120 ns, 5—150 ns

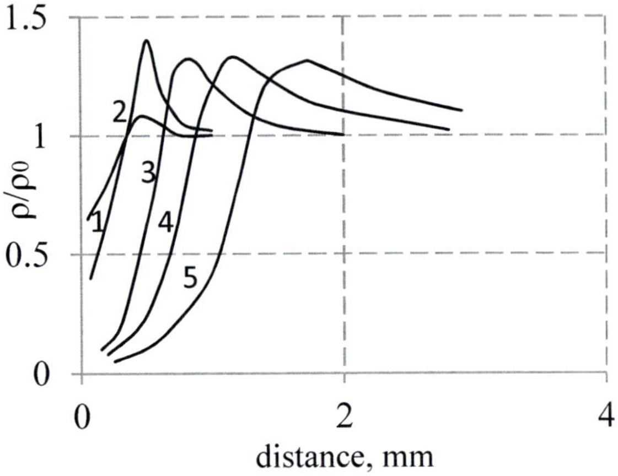

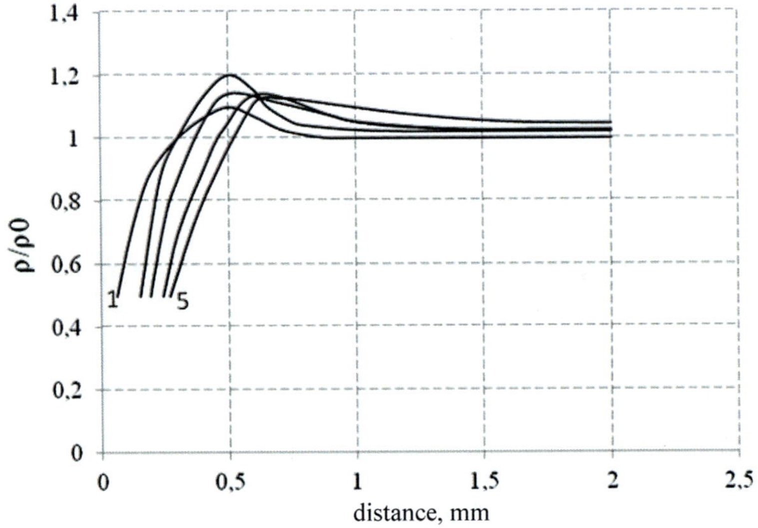

Fig. 3.2 Space profiles for $B = 3 \times 10^{-4}$ $n\textit{з}$ at the times: 1—45 ns, 2—60 ns, 3—90 ns, 4—120 ns, 5—150 ns

the maximum value of $\frac{\rho}{\rho_0} = 1.4$. Then, the structure begins to "wash-away", so by $t = 150$ ns; the dislocations are uniformly distributed in the medium. The lifetime of the mesoparticle is between 150 and 200 ns.

Increasing the damping coefficient to $B = 3 \times 10^{-4}$ ps leads to stopping the mesoparticle, which is localised near the load surface of the specimen.

The grouping of single-sign dislocations and the following wash-away can be considered short-time pulsations in dislocation density, resulting in velocity pulsations. In dynamic experiments, the pulsations are fixed in the form of the mesoparticle velocity distribution.

A mesoparticle as an ECD under conditions of high-velocity deformation is the short-lived accumulation of single-sign dislocations, the lifetime of which is on the order of 150–200 ns. In the process of mesoparticle motion, the pulsations of the mass velocity are generated, which are fixed in the form of the particle velocity distribution at the mesoscale.

3.3 Accounting for the Processes of Multiplication and Annihilation of Dislocations

Below, we present a detailed analytical solution to grouping the dislocation structure. In this case, the equation for dislocation density and flow has the form:

$$\frac{\partial \alpha}{\partial t} + \frac{\partial}{\partial x}(\alpha u_d) = J(\alpha u_d), \tag{3.6}$$

where α and u are the mean density and velocity of dislocations. The right-hand side of Eq. (3.6) can be taken in the form [3]:

$$\frac{\partial \alpha}{\partial t} + \frac{\partial}{\partial x}(\alpha u_d) = \alpha(\alpha u_d) - \beta(\alpha u_d)^2. \tag{3.7}$$

This form assumes that the multiplication and annihilation of dislocations depend only on the dislocation flow. As for the dislocation velocity, in the typical case, it also depends on the external and internal stresses as follows:

$$u_d = k\sigma^{\text{total}} = k\left(\sigma^{\text{ex}} + \sigma^{\text{int}}\right) \tag{3.8}$$

where σ^{ex} is the external stress and σ^{int} is the stress of the dislocation interactions. The latter can be taken in the form [4]:

$$u_d = k\left[\sigma(x) - D\int_{-\infty}^{\infty} \frac{\alpha(\xi, t)}{\xi - x} \mathrm{d}\xi\right], \tag{3.9}$$

Equation (3.7) has the stationary solutions:

$$F = 0;\ F = a/\beta;\ F = \frac{a}{B_0 \exp(-\alpha x) + \beta}, \tag{3.10}$$

where $F = \alpha u_d$. The problem is symmetrical so that we can solve for $x > 0$. The case $F = 0$ is not interesting, and $F = \alpha/\beta$ is asymptotic for $F = \frac{\alpha}{B_0 \exp(-\alpha x)+\beta}$ when $x \to \infty$. Multiplying Eq. (3.9) by α and expressing $\alpha(\xi)$ as a series of $(\xi - x)$, one obtains the general equation for α in the form:

$$F = k\sigma\alpha - 2DR\alpha\alpha'_x \tag{3.11}$$

Here, R is the screening radius for dislocation interactions. Consider the situation when the external stress changes as $\sigma = \sigma_0 + \sigma_1 \exp(-\delta x)$. Then the equation to be solved is the following:

$$\alpha\alpha'_x - \frac{\sigma_0 + \sigma_0 \exp(-\delta x)}{2DR} + \frac{\alpha}{2DRk} \frac{1}{B_0 \exp(-\alpha x) + \beta} = 0. \tag{3.12}$$

This is a first-order ordinary differential equation or with the boundary condition

$$\alpha = \alpha_0 \text{ at } x \to \infty. \tag{3.13}$$

As the dislocation flow $F = \frac{\alpha}{B_0 \exp(-\alpha x)+\beta}$ increases when $x \to \infty$ and $\sigma(x)$ decrease, the point x_* exists when $u_d(x_*) = k\sigma(x_*)$. Then the integral on the right-hand side of Eq. (3.9) vanishes and, in our approximation, σ'_x must vanish as well. The point x_* is an extreme point of the dislocation density function. For such a point to exist, the following inequality must be valid:

$$F(0) \le \sigma(0)\alpha(0).$$

If $\alpha'(0) = 0$ then $u(0) = k\sigma(0)$ and $\alpha(0) = \frac{\alpha}{(B_0+\beta)(\sigma_0+\sigma_1)}$.

However, another point, x_*, can exit where $\alpha'(x_*) = 0$. As this point is an external point, a region with a higher dislocation density must exist in the vicinity of the point. Thus, to have the entire picture of a stationary dislocation distribution, we must solve Eq. (3.12) with the boundary condition (3.13). But we are interested in the existence of a region with higher dislocation density and, therefore, the point x_* where $\alpha'(x_*) = 0$. Thus, we look for a solution of the form $\alpha(x) = \varphi_0 + \varphi_2(x - x_*)^2$. Representing $\sigma(x)$ and $F(x)$ as a series in terms of powers of $(x - x_x)$ and taking into account terms up to the second order, we can derive the following from Eq. (3.12)

$$L + M(x - x_x) + N(x - x_x)^2 = 0. \tag{3.14}$$

As (3.14) is valid for any x in the vicinity of x_* then

$$L = M = N = 0. \tag{3.15}$$

Here

$$L = (\sigma_0 + \sigma_1 \exp(-\delta x_*))(\beta + B_0 \exp(-\alpha x_*)\varphi_0 - \alpha/k; \tag{3.16}$$

$$N = -2\alpha B_0 \varphi_0 \varphi_2 \exp(-\alpha x_*) - A\varphi_2\big[(\sigma_0 + \sigma_1 \exp(-\delta x_*)\big])$$
$$\times \big[\beta + B_0 \exp(-\alpha x_*)\big] - A\varphi_0\big[(\sigma_0 + \sigma_1 \exp(-\delta x_*)\big]\frac{\alpha^2}{2} B_0 \exp(-\alpha x_*)$$
$$+ \sigma_1 B_0 \alpha \delta \exp(\alpha - \delta)x_* + \frac{\delta^2}{2}\sigma_1 \exp(-\delta x_*)(\beta + B_0 \exp(-\delta x_*))$$

and $A = \dfrac{1}{2DR}$

Equations (3.16) are the closed system of three algebraic equations with three unknown values φ_0, φ_2 and x_*. In this investigation, the system has been solved for two specific cases: (i) a high-stress gradient and (ii) a low-stress gradient, that is, for $\delta x_* \gg \alpha x_*$ and $\delta x_* \ll \alpha x_*$.

1. $\delta x_* \gg 1$.

In this case, the system (2.16) transforms into:

$$\sigma_0\big[\beta + \exp(-\alpha x_*)\big] = \frac{\alpha}{k\varphi_0};$$
$$2\varphi_0\big[\beta + \exp(-\alpha x_*)\big] + A\beta\sigma_0 B_0 \exp(-\alpha x_*) = 0;$$
$$2\varphi_0\varphi_2 \alpha B_0 \exp(-\alpha x_*) + A\varphi_2\sigma_0\big[\beta + B_0 \exp(-\alpha x_*)\big]$$
$$+ A\varphi_0\sigma_0 \frac{\alpha^2}{2} B_0 \exp(-\alpha x_*) = 0. \tag{3.17}$$

The solution to (3.17) has the form:

$$\varphi_o = \frac{\alpha}{4k\beta\sigma_0}\left(1 + \sqrt{\frac{8Lk\beta\sigma_0^2}{\alpha^2} + 1}\right); \tag{3.18}$$

$$\varphi_2 = \frac{1}{8} L\alpha\sigma_0 \left[\sqrt{\frac{8Lk\beta\sigma_0^2}{\alpha^2} + 1} - 3\right]. \tag{3.19}$$

As x_* is an external point of the function $\alpha(x)$, φ_2 must be negative, so the following inequality must be valid:

$$\sigma_0 < \frac{2DR}{k}\frac{\alpha^2}{\beta}. \tag{3.20}$$

In the case of (3.20), there exists a solution to Eq. (3.17), which allows the point x_* of maximum dislocation density to be determined. This dislocation group can be regarded as a particle at the next structural level of deformation.

2. $\alpha x_* \gg 1$.

In this case, the system (3.15) can be written as follows:

$$\begin{aligned}&\left[\sigma_0 + \sigma_1 \exp(-\delta x_*)\right]\beta\varphi_0 - \frac{\alpha}{k} = 0;\\&2\varphi_2 + L\delta\sigma_1 \exp(-\delta x_*) = 0;\\&\varphi_2\left[\sigma_0 + \sigma_1 \exp(-\delta x_*)\right] + \varphi_0 \frac{\delta^2}{2}\sigma_1 \exp(-\delta x_*) = 0.\end{aligned} \tag{3.21}$$

The solution to (3.21) has the form:

$$\varphi_0 = \sqrt{\frac{L\alpha}{k\beta\delta}}; \tag{3.22}$$

$$\varphi_2 = \frac{1}{2}\left(\delta^2\sqrt{\frac{L\alpha}{k\beta\delta}} - L\sigma_0\delta\right); \tag{3.23}$$

$$x_* = \frac{1}{\delta}\ln\frac{\sigma_1}{\sqrt{\frac{\alpha\delta}{Lk\delta}} - \sigma_0}. \tag{3.24}$$

This solution is possible under the condition

$$\sigma_0 < \sqrt{\frac{\alpha\delta}{Lk\beta}}. \tag{3.25}$$

As a result, it can be concluded that:

- The entire picture of dislocation distribution can be obtained through the solutions to Eqs. (3.12) with the boundary conditions (3.13).
- The two particular cases illustrated herein show that stable dislocation structure is possible only under the conditions (3.20) and (3.25). In the first case, both the loading stress and the coefficients α and β, which determine the evolution of dislocation structure, have to be taken into account. In the second case, the value of δ, which characterises the rate of change of the loading stress, determines the formation of the dislocation structure.
- The point x_* is an extreme point for the function $\alpha(x) = \varphi_0 + \varphi_2(x - x_*)$, where φ_0 and φ_2 are determined by Eqs. (3.24) and (3.25).

References

1. Meshcheryakov YuI, Prokuratova EI (1995) Kinetic theory of continuously distributed dislocations. Int J Solids Struct 32:1711–1726
2. Rozhdestvensky BL (1974) Picard's method as method for numerical solution to tasks of mathematical physics. Num Methods Continuum Mech 5(2):96–98. (Russian edition)
3. Prokuratova EI, Indeitzev DI (1995) The conditions for the dislocation distribution localization under dynamic loading. Dymat J Vl(¾):229–233
4. Head AK, Wood WW (1973) Dislocation group dynamics. Philos Mag 27:505–517

Chapter 4
Concept of the Mesoscale in Quasistatics and Dynamics

Abstract Under the conditions of dynamic straining, as distinct from the quasistatics, mesoparticles are not completely formed structures. The dynamic structure of the mesoscale is the characteristic of the *spatio-temporal process* in which the deformed medium is involved, but not characteristic of the *state of the medium*. The existence of a dynamic mesostructure is possible only within the time intervals during the loading. Outside of loading, the conditions for the existence of a dynamic mesoparticle disappear as the mesostructure attains the equilibrium state. In this situation, the dimensions of mesoparticles are determined by the particle velocity distribution and sensitively depend on the kind of material.

4.1 Introduction

The problem of an adequate description of macroscopic dynamic plasticity through ECD has been a subject of investigations for many years. A critical issue is that a direct transition from the atomic dislocation scale to the macroscale is unlikely. Specifically, to adequately describe the elastic precursor decay and elastic–plastic waves in solids, the initial density of dislocations must be increased by two or three orders of magnitude, which does not correspond to real values. This discrepancy suggests that the additive dislocation mechanisms of deformation are insufficient [1, 2]. Collective mechanisms need to be incorporated for the adequate modelling of dynamic plastic deformation. For the first step in this direction, at the beginning of the 1980s, the concept of the mesoscale was suggested—the elementary carriers of deformation are speculated to be the structures that occupy the intermediate position between the dislocation scale and the macroscale [3–8]. At the same time, the multi-scale mechanisms responsible for the behaviour of solids under dynamic loading are still purely understood. In this situation, the mesoscale phenomena for both aspects of deformation, quasistatics and dynamics, should be considered separately.

Y. Meshcheryakov, *Multiscale Mechanics of Shock Wave Processes*, Shock Wave and High Pressure Phenomena, https://doi.org/10.1007/978-981-16-4530-3_4

4.2 Quasistatics

Mesoscopic modelling of deformation and fracturing has been developed to resolve the contradictions inherent to traditional methods of description: on the one hand, the mechanics of a continuous medium, and, on the other hand, dislocation theory. The mechanics of the continuum medium describe the deformation processes without taking into account the non-uniformity of the internal structure of the material and determine the integral plastic and strength characteristics of the macroscopically uniform medium. Distinct from continuum mechanics, dislocation theory describes the microscopic behaviour of the deformed medium, which reveals the physical nature of structural non-uniformities. At the same time, numerous attempts to relate the physical theory of dislocations and continuum mechanics have been unsuccessful. At different stages of deformation, the physical processes, which have different activation energies and scales, play a role in leading mechanisms. There is a hierarchy of structural levels in a deformed crystal, making the description of deformation processes a multiscale problem that a single-level approach cannot solve.

Among all the structural levels, two endpoint levels have been determined. One endpoint is the microscopic (atomic) level h_2 or the scale of separate lattice defects. The specific scale of this level is $l_2 \leq (10^{-7} - 10^{-6})$cm. The basic physical processes herein are the kinetics of point defects, the reactions between individual dislocations and the diffusion plasticity. At the scale of l_2, the character of the microfractures is determined. The most important feature of the l_2 scale is the absence of collective interactions of dislocations because the specific volumes V_2 are too small. The basic objects of study at the microscopic level are the individual defects of the crystalline lattice. The commonly used mathematics herein are the methods of atomic physics, the concept of quasi-particles and simulation.

The second scale endpoint is the macroscopic level h_O with a specific scale of $l_0 \geq (10^{-3} - 10^{-1})$cm. The scale l_0 is the region of the continuum description of the ensemble of defects using distribution functions, such as the dislocation density $\rho(x)$. At the scale of h_O, the concept of the "point" for the effective elastic–plastic features inside the volume l_0^3 is introduced, which means that the properties of the individual dislocations are not essential. The basic mathematical apparatus for the study of the macroscale is the traditional apparatus of continuum elasticity, the theory of plasticity and creep.

Between these endpoints levels lies a wide region of structural scales, $l_1 \approx (10^{-6} - 10^{-3})$, of intermediate (mesoscopic) h_1 level. The specifics of h_1 structures include the following: (i) a dislocation group before an obstacle or a shear band, (ii) dislocation walls, (iii) microcracks, (vi) boundaries of fragments and (v) bands of rotational faults. The mathematical apparatus developed here has been absent up to now due to the complexity of the physical processes at this scale.

From the physical point of view, the main distinctive feature of the mesoscale is that in its specific volumes V_1 under the action of a stress field, the possibility for rebuilding the initially uniform and quasi-neutral distribution of dislocations into a non-uniform distribution appears, that is, heterogenisation of the dislocation

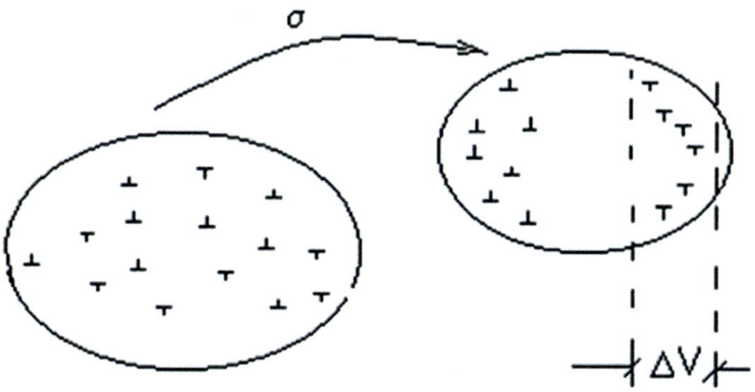

Fig. 4.1 Qualitative pattern of dislocation polarisation under applied stress

ensemble. If the specific volume of the structural heterogenisation equals the free run of dislocations, most of the dislocations from the mesovolumes V_1 are separated in sign and accumulated in narrow layers near the boundaries of the volumes. Thus, the mesovolumes (V_I) are regions that are essentially free from dislocations with high gradients of single-sign dislocation density within a boundary band, ΔV_1. It is clear that averaging over the dislocation ensemble of this condition is incorrect, and, hence, the continuum approach of macroscopic plasticity is not appropriate (Fig. 4.1).

The formation of regions with the above substructure leads to the appearance of the following specific features:

1. Large stress gradients, $\Delta\sigma = \sigma/l_{ms}$, achieve their maximum value at $l_{ms} = \lambda$, where λ is the free run of dislocations. Even a relatively small degree of dislocation polarisation, $\frac{\Delta\rho}{\rho} = 10^{-3}$, can create overstresses with a concentration coefficient of $\gamma \geq 3$. In the case of the full polarisation of the dislocation ensemble ($\Delta\rho = \rho$), the stress gradients are very high and may lead to activation of rotation modes of plasticity and the nucleation of microcracks.
2. The mesovolumes present a zone of plasticity around the materials in an elastic state (the straightened plastic deformation), and the grain boundary of V_1 is the elastic–plastic boundary. The large number of incomplete shears created at the scale $(1-10)\mu m$ drives the increased non-uniformity of the plastic deformation.
3. The mesovolumes with a polarised dislocation ensemble can be considered a quasi-particle with effective charge $\Delta q = \Delta\rho\Delta V_1$. Such mesovolumes have a larger long-range interaction radius compared to the initial quasi-neutral state where $\Delta\rho$ and Δq are small.
4. Thus, in plastically deformed crystals, in addition to atomic interactions, there are non-local interactions, which result from distributed dislocation charges and are different in size, kind and power.

Initially, it was accepted that the above features of mesostructures are present only in large plastic deformations. However, recent theoretical investigations in nonlinear mechanics have revealed the nucleation of heterogeneous structures in the elastic region of deformation [9]. Modern nonlinear microdeformation theory based on the solution to the sine–Gordon equation predicts the instability of a crystalline lattice subjected to large gradients of shear deformation in the nonlinear region of elastic deformation. In this case, these are the conditions for nucleation of mesorotations, shear bands and combinations thereof. These large-scale structures are nucleated well before the transition to macroscopic plasticity happens. If the deformation gradient does not exceed some critical value, the above structures disappear after the external stress is removed. Thus, at this stage, the process of nucleation of a mesostructure is reversible. The long-range interactions of mesoscale defects, if they are "frozen" into a crystal, can be considered as the interaction of the volumes containing these defects. In mesomechanics, these volumes are considered to be self-consistent carriers of deformation that define the mechanical behaviour of a solid.

4.3 Dynamics

In the quasistatic region of strain rates, the basic features of the mesoscale are well studied. At the same time, for high strain rates, the physical nature of the mesoscale remains unknown. Modern experimental methods still do not allow real-time investigation of the state of the inner structure of a material at the temporal and spatial scales of shock wave loading. In contrast, the microstructural investigations, which are commonly conducted after shock loading, cannot reveal the structural parameters of a material under high-velocity deformation because of the multiscale accommodation processes. Under the conditions of dynamic straining, as distinct from the quasistatics, mesoparticles are not completely formed structures. It is thought to be more natural to present mesoparticles as field structures transformed into well-known localised structures (shear bands, rotations, faults and combinations) only after a certain time interval. In dynamic deformation, the only feature of mesoparticles is that the particle velocity is identical for the mesovolume as a whole, independent of the kind of physical processes inside the mesovolume, including the motion of dislocations, twinning and the diffusion of point defects. Although these crystalline structure defects are responsible for the deformation of mesoparticles, after formation is complete, the mesoparticle can be presented as a structureless object that moves as a whole together with analogous objects during dynamic straining. In this situation, the inertial properties are the important features of mesoparticles.

It follows from the above consideration that the dynamic structure of the mesoscale is characteristic of the *spatio-temporal process* in which the deformed medium is involved, but not characteristic of the *state of the medium*. The existence of a dynamic mesostructure is possible only within the time intervals during the loading. Outside of loading, the conditions for the existence of a dynamic mesoparticle disappear as the mesostructure attains the equilibrium state. In this state, the difference in the velocities of neighbouring mesoparticles disappears as well as the boundaries between neighbouring mesoparticles. The only possibility for the visualisation of the mesostructure is the creation of the conditions at which the mesoparticles are instantly removed from one another, so there is no time for their velocities to equalise.

This situation is realised within the spall zone of shock-deformed material, where for a very short time, comprehensive tension stress is initiated. During spallation, the spall surfaces of the target are quickly removed from each other, conserving the structure of the material inside the spall zone. Accordingly, the boundaries between the mesoparticles which separate the volumes of the medium with different velocities are also conserved. Under conditions of high-velocity tension, the separation of mesoparticles happens as the mesoparticle themselves transforms into grains.

Some typical relict mesostructures are presented in Fig. 4.2a–h. This powder has been found in the spall zone of copper targets after shock loading. The first type of mesostructure formed has an amorphous structure, which is nucleated during the nanosecond duration of the pressure pulse. The second kind of mesostructure formed has a crystallised structure, which nucleated during a stress pulse of hundreds of nanoseconds. In the second case, each mesoparticle contains a series of shear bands spaced by 5–10 nm. In copper, these shear bands provide a homogenisation of the inner structure of the mesoparticle. As a rule, the motion of the medium during this kind of homogenisation is characterised by the velocity distribution. The velocity distribution is fixed in the form of the particle velocity dispersion by the velocity interference technique (see below).

Post-shock microstructural investigations show that the dynamically deformed medium may include several mesostructural sub-scales depending on the strain rate. For example, as presented in Fig. 4.3, the spatial distributions of the horizontal (d and D) and the vertical (h and H) pieces of spall fracture for 38CrNi3MoV steel reflect the two sub-levels of dynamic fracture. Spallation is seen to occur here in two scales simultaneously: at the mesoscale 1 (1–10) μm and the mesoscale 2 (50–500) μm. At the same time, Fig. 4.4 presents post-shock structures which show spallation only at the mesoscale 2 scale.

One distinctive feature of dynamic mesoparticles as the carriers of deformation at the mesoscale is the velocity distribution. The kinetic energy transferred to the deformed solid from the external load is distributed between the mean (flow) motion of the medium and the chaotic motion of the mesovolumes.

In Fig. 4.5, three random configurations of a shock front in a heterogeneous medium are qualitatively shown. The dotted lines represent the mean positions of the front in a uniform medium, whereas the meandering lines are the random positions of separate pieces of the shock front. Here, the small steps are the particles of mesoscale 1, whereas the large steps are the particles of mesoscale 2. In this scenario, one can regard the motion of the shock front in a heterogeneous medium as the superposition of two modes of motion: a mean motion, which is approximately the motion of the plane front, and rapidly fluctuating motions, which are due to the action of local stress fields.

Such stochastic processes can be characterised with the particle velocity distribution function (PVDF) or its statistical moments. The zero statistical moment of the PVDF, $\rho(r, t)$, is the density; the first statistical moment is the mathematical expectation or the mean particle velocity, and the second moment is the particle velocity dispersion.

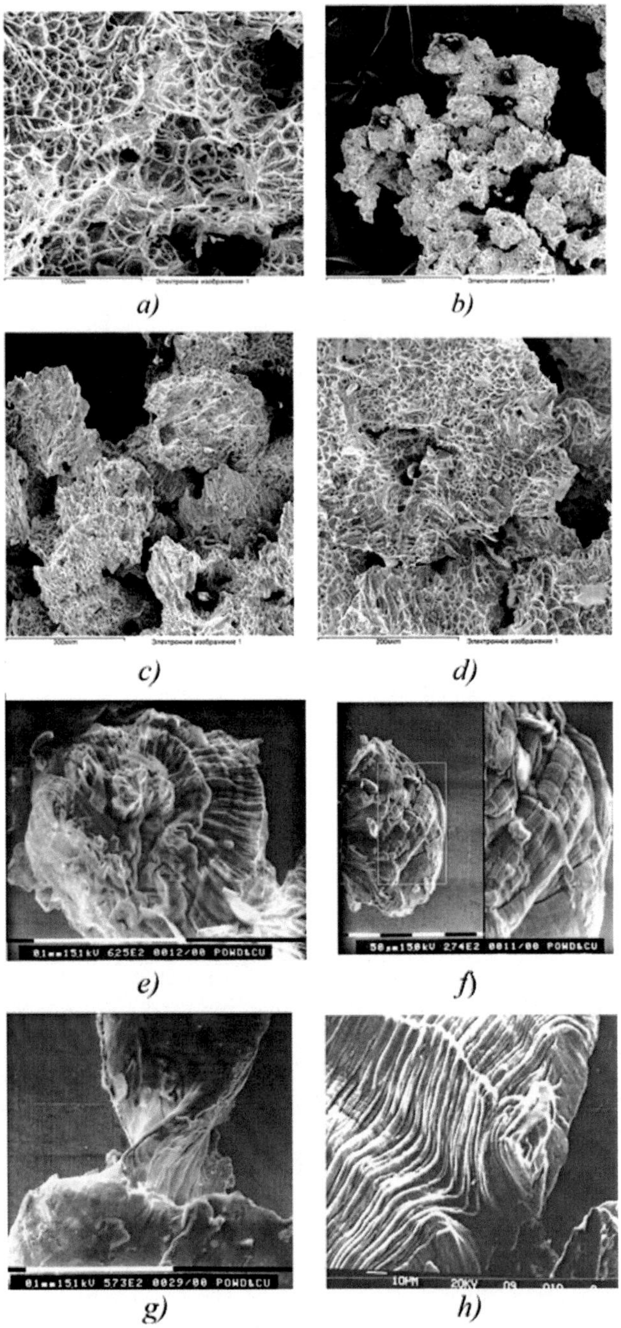

Fig. 4.2 Relict mesoparticles from the spall zone of shock-deformed copper: (**a–d**)—amorphous formations from the pulse duration of ~30–50 ns, (**e–h**)—crystallised formations with shear bands

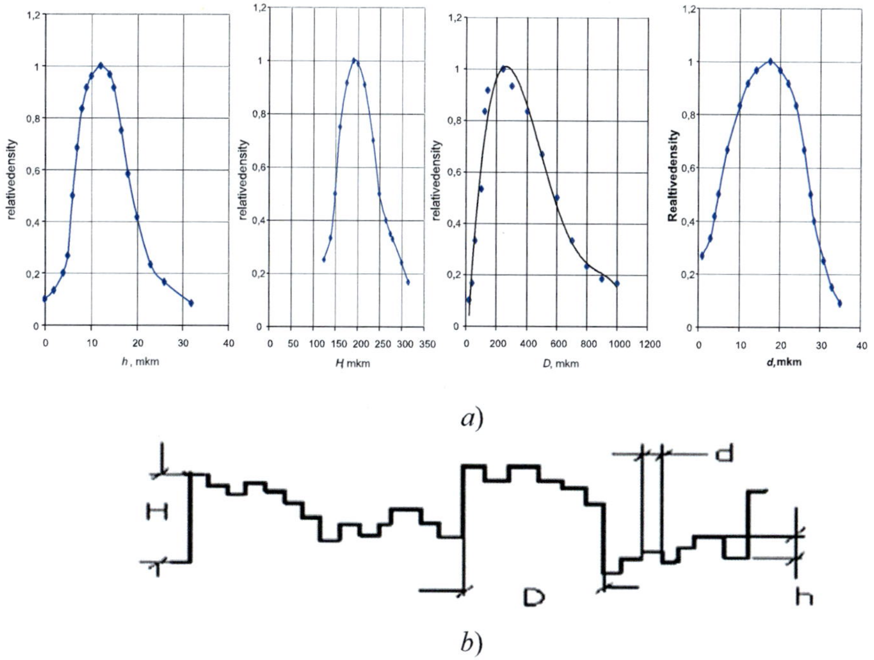

Fig. 4.3 Spatial distributions of pieces of spall fracture in 38CrNi3MoV steel

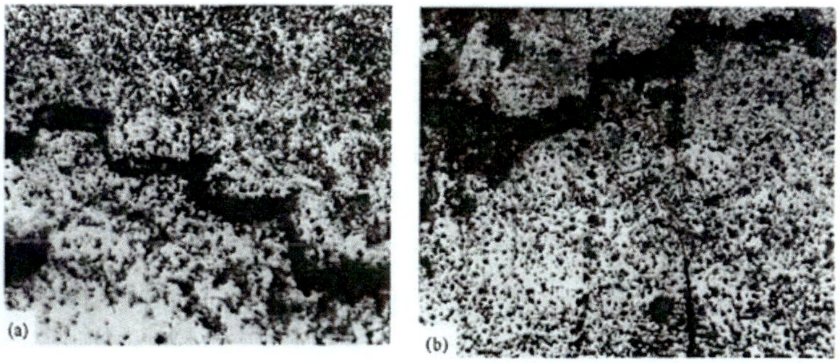

Fig. 4.4 Post-shock structures of spall cracks in 30CrNi4Mo steel

$$\rho(r,t) = \int_{-\infty}^{\infty} f(r,v,t)\mathrm{d}v \tag{4.1}$$

$$u(x,t) = \frac{\int_{-\infty}^{\infty} v f(x,v,t)\mathrm{d}v}{\rho} \tag{4.2}$$

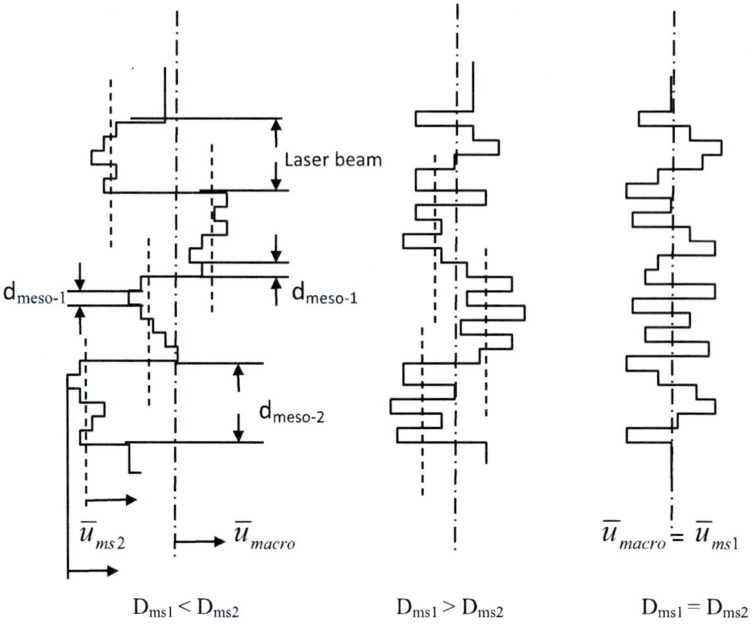

Fig. 4.5 Qualitative pattern of *U-L* velocity space configurations of the shock front for different relations between velocity dispersions at mesoscale 1 and mesoscale 2

$$D^2 = \int_{-\infty}^{\infty} (v - \overline{v})^2 f(x, v, t) dv, \tag{4.3}$$

Typically, the non-equilibrium PVDF is asymmetrical. The asymmetry of the PVDF can be characterised using Pearson's asymmetry as follows:

$$s = \frac{\overline{u} - v_{\max}}{D}, \tag{4.4}$$

where $\overline{u}$ is the mean value of the particle velocity and $v_{\max}$ is the particle velocity corresponding to the maximum PVDF. In the process of dynamic straining, the sub-levels of mesostructure discussed previously can have separate velocity distributions. Depending on the intensity of each distribution, the following situations may occur, as shown in Fig. 4.5:

(1). The velocity distribution at mesoscale 1 is smaller than that at mesoscale 2.
(2). The velocity distribution at mesoscale 1 is greater than that at mesoscale 2.
(3). The velocity distributions at the mesoscale 1 and mesoscale 2 are identical.

It has been shown experimentally that the mesoparticle velocity distribution function varies along the average velocity profile, u(t). The maximum dispersion occurs

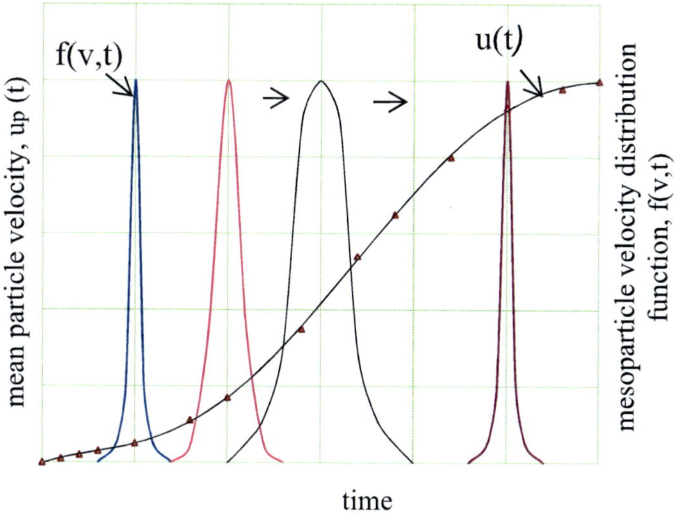

Fig. 4.6 Qualitative picture of the evolution of the velocity distribution function over the unsteady shock front

in the middle of the steady plastic wave front. The dispersion behaviour is not evident for unsteady shock fronts. It may increase along the plastic front up to the plateau of the compressive pulse. Qualitative pictures of the mesoparticle velocity distribution behaviour at the mesoscale1 for steady shock front is shown in Fig. 4.6. In each case, the rate of Maxwellisation of the velocity distribution of mesoscale 1 is much higher than the rate of change of the mean velocity at the plastic front. Therefore, for most materials, the velocity distribution of the mesoscale 1 in steady shock waves is in equilibrium, and it has a symmetrical (Maxwellian) shape, although its width changes along the plastic front.

References

1. Johnson JN, Jones OE, Michaels TE (1970) Dislocation dynamics and single-crystal constitutive relation: shock-wave propagation and precursor decay. J Appl Phys 41(6):2230–2239
2. Meshcheryakov YI (2016) Particle velocity non-uniformity and steady-wave propagation. Shock Waves Detonations Explos 26(3). https://doi.org/10.1007/S0093-016-0659-7
3. Panin VE, Egorushkin VE, Panin AV (2006) Physical Mesomechanics of deformed solid as multiscale system. Phys Mesomech 9(3):9–22
4. Makarov PV (2004) About hierarchic nature of deformation and fracture of solids. Phys Mesomech 7(4):25–34
5. Rybin VV (1986) Large plastic deformations. M. Metallurgy, 224 p
6. Vladimirov VI, Ivanov VN, Priemskii NM (1986) Mesoscopical level of deformation. In: Zhurkov SI, Nauka M (eds) Physics of strength and plasticity, pp 69–80
7. Baer MR, Trott WM (2001) Mesoscale description of shock loade heterogeneous porous materials. Bull Am Phys Soc 46:102

8. . Baer MR, Computatinal modeling of heterogeneous reactive materials at the mesoscale. In: Furnish MD, Chhabildas LC, Nixon RS (eds) Shock compression of condenced matter-1999. The APS Proceedings, New York, pp27–33.
9. Aero EL (2000) Microscale deformations in duble scale lattice transitions under crytical shear. Phys Solid 42(6):1147–1153

Chapter 5
The Mesoscale Velocity Distribution and Change of Regime of Shock Wave Propagation

Abstract We consider the propagation of a plane elastic–plastic wave in a heterogeneous medium where the dynamic behaviour of the medium is determined by the high statistical moments of the velocity distribution function—the mesoparticle velocity dispersion and the asymmetry of the mesoparticle distribution function. In the case of an unsteady plastic wave, the rate of the plastic deformation u^{pl} is presented as the sum of the equilibrium and non-equilibrium parts: $u^{pl} = u_p^{pl} + u_{\mathcal{H}}^{pl}$. The regime for the shock wave propagation changes when the rate of change of the mesoparticle velocity dispersion becomes higher the rate of change of the mean particle velocity. Physically, this means that the wave motion transits to a decaying regime. Thus, the condition for breaking the plastic front is determined by the ratio of the velocity variance and mean velocity and by the ratio of their rates of change. This means that in the dynamic processes, the rate of change of the particle velocity dispersion is the basic control parameter of the shock wave process. The obtained criterion has been applied to real experiments with the shock loaded materials, such as copper, aluminium alloys and steels. In experiments, the temporal behaviour of the velocity dispersion and mean particle velocity can be registered in real time in the processes of shock loading by using the interference technique. Shock tests of D16 aluminium alloy reveal the following features of dynamic response to shock loading: (1) The region of strain rates corresponding to impact velocities of 85–451 m/s is subdivided into two sub-regions at the impact velocity of 382 m/s. In the first sub-region (85–382 m/s), the velocity defect remains constant at ~20 m/s. In the second sub-region, corresponding to within impact velocity region of 382-451 m/s, the velocity defect grows with increasing impact velocity. This region corresponds to the irreversible regime of interscale momentum and energy exchange. (2) At the impact velocity of 382 m/s, the velocity variance equals the velocity defects, corresponding to the equality of the local strain rate and the macroscopic strain rate, just as that impact velocity corresponds to maximum spall strength of D16 Al alloy.

Y. Meshcheryakov, *Multiscale Mechanics of Shock Wave Processes*, Shock Wave and High Pressure Phenomena, https://doi.org/10.1007/978-981-16-4530-3_5

5.1 Introduction

In the case of the shock compression of a solid, the most evident indication of a change to the regime of dynamic deformation is thought to be the change of the regime of the shock wave propagation. This change occurs due to the complex behaviour of the material structure at several scales. In a steady shock wave, the exchange of momentum between scale levels is reversible. To change the regime, the process of dynamic deformation must be non-steady. To determine the threshold regimes of energy and momentum exchange between scales, a theoretical approach must incorporate the non-equilibrium mechanisms of dynamic deformation. In continuum mechanics, the description of the shock wave processes is based on locking the balance equations with the equations of state. In the present chapter, to describe the non-equilibrium processes at the mesoscale and determine the threshold regimes of dynamic deformation, the locking of the balance equations for shock wave propagation is realised through the non-equilibrium mesoparticle velocity distribution function. We consider the propagation of a plane elastic–plastic wave in a heterogeneous medium where the dynamic behaviour of the medium is determined not only by the mean particle velocity (as in continuum mechanics) but also by the high statistical moments of the velocity distribution function—the mesoparticle velocity dispersion and the asymmetry of the mesoparticle distribution function.

5.2 The Change of Regime of Shock Wave Propagation

Without taking the thermo-conductivity processes into account, the motion of a dynamically deformed medium is described with two balance equations:

$$\rho \frac{\partial u}{\partial t} = \frac{\partial \sigma_1}{\partial x}, \tag{5.1}$$

$$\frac{\partial u}{\partial x} = \frac{\partial \varepsilon}{\partial t}; \tag{5.2}$$

where u is the particle velocity, σ_1 is the normal component of stress, and ε and $\frac{\partial \varepsilon}{\partial t}$ are the total (elastic plus plastic) deformation and the strain rate in the propagation direction of the wave. The strain rate can be presented in the form:

$$\frac{\partial \varepsilon}{\partial t} = \frac{\partial \varepsilon^e}{\partial t} + \frac{\partial u^{pl}}{\partial x}. \tag{5.3}$$

Then, Eq. (5.2) can be written as follows:

$$\frac{\partial u}{\partial x} = \frac{\partial \varepsilon^e}{\partial t} + \frac{\partial u^{pl}}{\partial x}, \tag{5.4}$$

where

$$\varepsilon_1^e = (\lambda + 2\mu)^{-1}\sigma_1^e \tag{5.5}$$

is the one-dimensional elastic deformation, and λ and μ are the Lamé constants. The normal stress component σ_1 can be written using the spherical component (pressure) P and the deviator S_1.

$$\sigma_1 = P - S_1. \tag{5.6}$$

Equations (5.1)–(5.4) can be reduced to the following second-order differential equation:

$$\frac{\partial^2 u}{\partial x^2} - \frac{1}{C_1^2}\frac{\partial^2 u}{\partial t^2} = \frac{\partial^2 u^{pl}}{\partial x^2} - \frac{1}{(\lambda + 2\mu)}\frac{\partial^2 \sigma_1}{\partial x \partial t} \tag{5.7}$$

where $C_1 = \left(\frac{\lambda+2\mu}{\rho}\right)^{\frac{1}{2}}$ is the longitudinal sound velocity.

We consider the right-hand side of Eq. (5.7). In the case of an unsteady plastic wave, the velocity of the plastic deformation u^{pl} can be presented as the sum of the equilibrium and non-equilibrium parts:

$$u^{pl} = u_p^{pl} + u_{\mathcal{H}}^{pl}, \tag{5.8}$$

The mesostructures have two very important features:

1. The mesostructure is the totality of the polarised sign dislocation charges.
2. The interaction of such mesoparticles is long range.

Both features are analogous to the characteristics of charged particles of plasma. We use this analogy in order to establish the coupling between the macroscopic response of a solid to dynamic loading and the kinetics of a mesostructure. Under quasi-equilibrium conditions, the behaviour of an ensemble of particles with random interaction character is described by the Fokker–Planck equation for the particle velocity distribution function:

$$\frac{\partial f}{\partial t} + \frac{\partial f}{\partial r} + \frac{F}{m}\frac{\partial f}{\partial u} = -\frac{\partial}{\partial u}(F_1 f) + \frac{1}{2}\frac{\partial^2}{\partial u^2}(F_2 f). \tag{5.9}$$

The left-hand side of this equation describes the convective transportation of the probability density, whereas the right-hand side shows how this transportation occurs. The first term in the right-hand side of the equation, the so-called drift term, characterises the mean change of the probability density in the velocity space, whereas the second term describes the influence of random velocity fluctuations. The coefficients F_1 and F_2 are the diffusion coefficients of Fokker–Planck equation:

1. $F_1 = \frac{\langle \Delta u_1 \rangle}{\Delta t}$ is the first diffusion coefficient or the coefficient of dynamic friction.

2. $F_2 = \frac{\langle \Delta u_2 \Delta u_2 \rangle}{\Delta t}$ is the second diffusion coefficient, which is determined by the rate of change of the velocity dispersion.

The difference between the values $\langle \Delta u_1 \rangle$ and $\langle \Delta u_2 \rangle$ follows from the assumption introduced by Hubbard [1] for the ensemble of charged particles that experience long-range interactions. By analogy with the ensemble of charged particles, the motion of the mesoparticles in random stress fields may be presented in the form of a superposition of two modes of motion: Mode 1 is the motion of the mesoparticle for the time of correlation, and Mode 2 is the quickly fluctuating motion of mesoparticle velocities because of the actions of the random stresses within the time interval between correlations. Under this assumption, $\langle \Delta u_1 \rangle$ is the mean deviation of the mesoparticle velocity, which is small compared to the macroscopic changes of the velocity, for example, for the time duration of the plastic front of the shock wave. However, this change is sufficiently large compared to the change of mesoparticle velocity caused by the interaction of the mesoparticles with each other. As Mode 1 is the motion of the mesoparticle for the correlation time, this motion flows with the velocity u, which is determined by the action of the stress $\sigma(u\tau, \tau)$, where τ is the duration of the correlation process. In this case, the change of the velocity is determined by the integral:

$$\Delta u = \frac{s}{m} \int\limits_0^{\Delta t} \sigma(u\tau, \tau) \mathrm{d}\tau \tag{5.10}$$

Here, $\sigma(u\tau, \tau)$ is the stress determined by the value of the displacement of the mesoparticle for the correlation time, s is the square of the cross section of the mesoparticle, and m is the mass of the mesoparticle.

In a dynamically deformed medium, the diffusion coefficients F_1 and F_2 are not independent values. In order to show this dependence, we use the formalism of the correlation functions developed in [1, 1]. As was previously indicated, the justification for the application of this approach is the long-range interaction of the mesoparticles. To determine the second diffusion coefficient, one should account for the motion of the mesoparticle in Mode 2; that is, only random deviations of the particle trajectory between successive correlations are considered. For this time interval, the mesoparticle is subjected to random stress fields such that its motion occurs along the random trajectory:

$$r(t) = u\tau + \int\limits_0^{\Delta t} \sigma\left(u\tau'', \tau''\right) \mathrm{d}\tau''.$$

In the case of the motion of the shock front in a non-uniform medium, the position of the mesoparticle is determined by two components. The first component describes the mean displacement of the mesoparticle $u\tau$ during the correlation interaction of the mesoparticles with each other. The second component characterises the random deviation of the mesoparticle because of the random stress fields at the mesoscale. In this case, the total stress acting on the mesoparticle equals:

$$[\sigma(r\tau,\tau)] = \sigma(u\tau,\tau) + \frac{s}{m}\int\limits_0^{\tau'} d\tau'\sigma\left(u\tau'',\tau''\right)\frac{\partial}{\partial r}\sigma(u\tau,\tau). \tag{5.11}$$

where the Taylor expansion is applied. The mean change of mesoparticle velocity equals:

$$\left[\frac{\langle\Delta u\rangle}{\Delta t}\right] = \frac{s^2}{m^2}\int\limits_0^{\Delta t} d\tau\int\limits_0^{\tau} d\tau'\int\limits_0^{\tau'} d\tau''\left\langle\sigma(u\tau'',\tau''\frac{\partial}{\partial r}\sigma(u\tau,\tau)\right\rangle, \tag{5.12}$$

Changing the integration limits and taking into account that this integral depends only on the difference τ-τ'', one obtains:

$$\left[\frac{\langle\Delta u\rangle}{\Delta t}\right] = \frac{s^2}{m^2}\int\limits_{-\infty}^{\infty}\tau\left\langle\sigma(u\tau'',\tau''\frac{\partial}{\partial r}\sigma(u\tau,\tau)\right\rangle d\tau. \tag{5.13}$$

Herein, the assumption is introduced that the time interval Δt is much greater than the time interval of the correlations. It has also been taken into account that

$$\langle\sigma(\upsilon\tau,\tau)\rangle = 0. \tag{5.14}$$

In the determination of the second diffusion coefficient, the values of the stresses for different moments of the time interval τ and the relative displacements $u\tau$ are taken into account:

$$\left[\frac{\langle\Delta u\Delta u\rangle}{\Delta t}\right] = \frac{s^2}{m^2}\int\limits_{-\infty}^{\infty}\langle\sigma(0,0)\sigma(u\tau,\tau)\rangle d\tau. \tag{5.15}$$

It has been assumed that the time interval between correlations is much longer than the duration of the correlations themselves. Comparing Eqs. (5.13) and (5.12) yields:

$$\frac{\langle\Delta u_1\rangle}{\Delta t} = \frac{1}{2}\frac{\partial}{\partial u}\left[\frac{\langle\Delta u\Delta u\rangle}{\Delta t}\right]. \tag{5.16}$$

It follows from this equation that the first diffusion coefficient can be expressed through mesoparticle velocity dispersion; that is, it is determined by the rate of change of the velocity dispersion. The time interval Δt can be considered the relaxation time of the mesoparticle velocity distribution function τ_r. Then, the first diffusion coefficient can be determined as the rate of change of the mean particle velocity for the relaxation time of the mesostructure to the equilibrium state:

$$F_1 = \frac{\langle \Delta u_1 \rangle}{\Delta t} = \frac{\langle \Delta u \rangle}{\tau_r}. \tag{5.17}$$

Accordingly, the second diffusion coefficient equals:

$$F_2 = \frac{\langle \Delta u_2 \Delta u_2 \rangle}{\Delta t} = \frac{\langle \Delta u_2 \Delta u_2 \rangle}{\tau_r} \tag{5.18}$$

Equation (5.16) can be written in the form:

$$\frac{\langle \Delta u_1 \rangle}{\tau_r} = \frac{1}{2}\frac{1}{\tau_r}\frac{\partial \langle \Delta u_2 \Delta u_2 \rangle}{\partial u} \tag{5.19}$$

or

$$\langle \Delta u_1 \rangle = \frac{1}{2}\frac{\partial \langle \Delta u_2 \Delta u_2 \rangle}{\partial u}. \tag{5.20}$$

Equation (5.20) determines the relaxation of the mesoparticle distribution function to equilibrium. The relationship obtained suggests that the rate of non-equilibrium mass transportation is determined by the rate of change of the velocity dispersion of mesoparticles: that is,

$$u_{ne}^{pl} = \frac{1}{2}\frac{\partial \langle \Delta u_2 \Delta u_2 \rangle}{\partial u}. \tag{5.21}$$

Let us consider the second term in the left-hand side of Eq. (5.7). Typically, the stress also consists of equilibrium and non-equilibrium components:

$$\sigma = \sigma_{\text{eq}} + \sigma_{\text{ne}}, \tag{5.22}$$

where σ_{eq} characterises the stress at the Hugoniot adiabate:

$$\sigma_{\text{eq}} = \rho C u_{\text{eq}}. \tag{5.23}$$

The stress tensor, $\sigma_{\text{ne}} = P - S$, includes the spherical component P and the deviator S. The latter determines the relaxation features of the medium. It is assumed that the deviator does not include non-equilibrium momentum transportation, that is, $S = 0$. The momentum transportation in a shock wave is realised by pumping over the fluctuative component of the mesoparticle velocity from macroscale to mesoscale. The velocity fluctuations at the mesoscale play the same role as the thermal fluctuations at the microscale: they determine the mean pressure in the medium. Therefore, the spherical stress component can be determined from the mesoparticle velocity dispersion or velocity variance:

$$P = \rho C D. \tag{5.24}$$

Substitution of Eqs. (5.22) and (5.23) into Eq. (5.7) yields:

$$C_0 \frac{\partial^2 u}{\partial x^2} - \frac{\partial^2 u}{\partial t^2} = C_0 \frac{\partial^2 u^{pl}}{\partial x^2} + \frac{1}{2} C_0 \frac{\partial^2}{\partial x^2} \frac{\partial D^2}{\partial u} - \frac{\partial^2}{\partial x \partial t}\left(C_0 u^{pl}\right) - \frac{\partial}{\partial t}\left(\frac{\partial D^2}{\partial x}\right). \tag{5.25}$$

The first and second terms in the right-hand side of Eq. (5.25) which relate the equilibrium components of the momentum transport mutually disappear, as at the Hugoniot adiabate, the process of dynamic straining is in equilibrium. In other words, the equilibrium components of the mass transportation rate u_{eq} do not incorporate the change in the integral velocity profile of the shock wave at the mesoscale. Then, Eq. (5.25) can be written in the form:^

$$C_0 \frac{\partial^2 u}{\partial x^2} - \frac{\partial^2 u}{\partial t^2} = C_0 \frac{\partial^2}{\partial x^2} \frac{\partial D^2}{\partial u} - \frac{\partial}{\partial t}\left(\frac{\partial D^2}{\partial x}\right). \tag{5.26}$$

The terms in the right part of Eq. (5.26) can be expressed with the following relationships:

$$\frac{\partial}{\partial t}\left[\frac{\partial (D^2)}{\partial x}\right] = 2 \frac{\partial}{\partial t}\left[u \frac{D}{u} \frac{\partial D}{\partial x}\right] = 2 \frac{\partial u}{\partial t}\left[\frac{D}{u} \frac{\partial D}{\partial x}\right] + 2u \frac{\partial}{\partial t}\left[\frac{D}{u} \frac{\partial D}{\partial x}\right]; \tag{5.27}$$

$$\frac{\partial}{\partial x}\left[\frac{\partial (D^2)}{\partial u}\right] = 2 \frac{\partial}{\partial x}\left[u \frac{D}{u} \frac{\partial D}{\partial u}\right] - 2 \frac{\partial u}{\partial x}\left[\frac{D}{u} \frac{\partial D}{\partial u}\right] + 2u \frac{\partial}{\partial x}\left[\frac{D}{u} \frac{\partial D}{\partial u}\right]; \tag{5.28}$$

$$\frac{\partial^2}{\partial x^2}\left[\frac{\partial (D^2)}{\partial u}\right] = 2 \frac{\partial^2}{\partial x^2}\left[u \frac{D}{u} \frac{\partial D}{\partial u}\right] + 4 \frac{\partial u}{\partial x} \frac{\partial}{\partial x}\left[\frac{D}{u} \frac{\partial D}{\partial u}\right] + 2u \frac{\partial^2}{\partial x^2}\left[\frac{D}{u} \frac{\partial D}{\partial u}\right]. \tag{5.29}$$

Using Eqs. (5.27)–(5.29), Eq. (5.26) takes the form:

$$C_0\left[1 - \left(\frac{D}{u} \frac{\partial D}{\partial u}\right)\right] \frac{\partial^2}{\partial x^2} - \frac{\partial^2 u}{\partial t^2} = 2 \frac{\partial^2}{\partial x^2}\left[u \frac{D}{u} \frac{\partial D}{\partial u}\right] + 4 \frac{\partial u}{\partial x} \frac{\partial}{\partial x}\left[\frac{D}{u} \frac{\partial D}{\partial u}\right] + 2u \frac{\partial^2}{\partial x^2}\left[\frac{D}{u} \frac{\partial D}{\partial u}\right] - 2u \frac{\partial}{\partial t}\left[\frac{D}{u} \frac{\partial D}{\partial x}\right] - 2 \frac{\partial u}{\mathbb{R}t}\left[\frac{D}{u} \frac{\partial D}{\partial x}\right]; \tag{5.30}$$

when

$$\left(\frac{D}{u} \frac{\partial D}{\partial u}\right) = 1, \tag{5.31}$$

Equation (5.30) is reduced to the equation:

$$\frac{\partial^2 u}{\partial t^2} = -\frac{\partial u}{\partial t}\left(\frac{D}{u}\frac{\partial D}{\partial x}\right). \tag{5.32}$$

The expression $\left(\frac{D}{u}\frac{\partial D}{\partial x}\right)$ can be transformed into

$$\left(\frac{D}{u}\frac{\partial D}{\partial x}\right) = \frac{D}{u}\frac{\partial D}{\partial u}\frac{\partial u}{\partial x},$$

and taking Eq. (5.32) into account, one obtains:

$$\left(\frac{D}{u}\frac{\partial D}{\partial x}\right) = \frac{1}{C_0}\frac{\partial u}{\partial t}. \tag{5.33}$$

Substitution of Eq. (5.39) into (5.5) yields:

$$\frac{\partial^2 u}{\partial t^2} + \frac{1}{C_0}\left(\frac{\partial u}{\partial t}\right)^2 = 0. \tag{5.34}$$

The solution to this equation has the form:

$$u = u_0 + u_1 lnt. \tag{5.35}$$

The relationship (5.31) can be reduced to

$$\left(\frac{D}{u}\frac{\partial D}{\partial u}\right) = \left(\frac{D}{u}\frac{\partial D/\partial t}{\partial u/\partial t}\right) = 1. \tag{5.36}$$

When Eq. (5.36) is obtained, the form of Eq. (5.30) changes. Physically, this means that the wave motion transits to a decaying regime. Thus, the condition for breaking the plastic front is determined by the ratio of the velocity variance and mean velocity and by the ratio of their rates of change. An analogous relationship has been obtained based on statistical non-local transport theory in [3].

Now let us estimate the above criteria for the experimental free surface velocity profiles presented in Fig. 5.1. The velocity profiles are obtained for 30CrNi4Mo armour steel targets loaded with the uniaxial strain conditions of shock loading. For the steepest part of the plastic front, *BB'*, the rate of change of the mean particle velocity equals:

$$\frac{\mathrm{d}u}{\mathrm{d}t} = \frac{25 \times 10^2 (\mathrm{cm/s})}{(344 - 320) \times 10^{-9} (\mathrm{s})} = 3.14 \times 10^{11}\ \mathrm{cm/s^2}.$$

For the same time interval, the rate of change of the velocity variance equals:

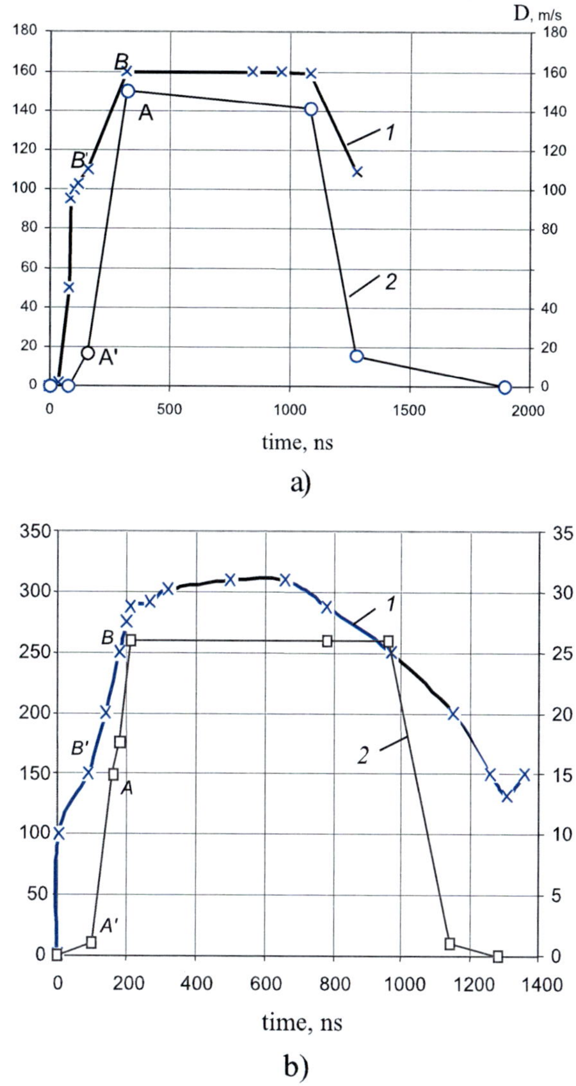

Fig. 5.1 Time-resolved free surface profiles (1) and velocity variance profiles (2) for **a** the first and **b** second sets of 30CrNi4Mo steel targets

$$\frac{\mathrm{d}D}{\mathrm{d}t} = \frac{25.9 \times 10^2\,(\mathrm{cm/s})}{(344 - 320) \times 10^{-9}(\mathrm{s})} = 0.15 \times 10^{11}\,\mathrm{cm/s}^2.$$

The mean particle velocity u and the velocity variance D at the points B and A equal: $u = 290\,\mathrm{m/s}$ and $D = 25.9\,\mathrm{m/s}$. Then, $\frac{\dot{D}}{\dot{u}} = 0.047$, $\frac{D}{u} = 0.07$ and $\left(\frac{D}{u}\frac{\dot{D}}{\dot{u}}\right) = 0.003 << 1$.

This result corresponds to the evolutional regime of momentum exchange. Analogous calculations for the mean particle velocity and velocity variance presented in Fig. 5.1 produced: $U = 160$ m/s, $D = 140$ m/s, $\dot{D} = 0.75 \times 10^{11}$(m/s) and $\left(\frac{D}{u}\frac{\dot{D}}{\dot{u}}\right) = 2.18 > 1$, which correspond to a catastrophic scenario for the exchange of macro- and meso-energy. In this case, at point B, the plastic front breaks, and the resulting free surface velocity at the plateau of the compressive pulse is much smaller. This decrease in the mean particle velocity indicates the momentum exchange between the macroscale and the mesoscale. In this specific case, the decrease ("defect") of the particle velocity equals $\Delta U_{\text{def}} = U_{\text{imp}} - U_{\text{fs}} = 320 - 160$ m/s. Thus, the different thermo-mechanical treatments of steel, which create the structural non-uniformity of the material, lead to a different strength behaviour of the material under shock loading.

5.3 Irreversible Momentum Exchange

As discussed above, for the turbulent regime of flow to be initiated, the solid must transit into the structurally unstable state. This transition is accompanied by localised deformation. After the point of unstable flow is passed, the heterogeneous deformation is initiated. The transition of the solid into the structurally unstable state is statistical in nature, following from the criterion (5.36), which shows that the ratio plays the key role in the transition of the material into the structurally unstable state $(\mathrm{d}D/\mathrm{d}t)/(\mathrm{d}u/\mathrm{d}t)$. The transition into the structurally unstable state happens when the rate of change of the velocity variation becomes higher than the rate of change of the mean particle velocity.

Below, the criterion (5.36) is applied to the particle velocity profiles for D16 aluminium alloy that has been shocked under uniaxial strain conditions within the impact velocity region of 85–450 m/s. The first velocity profile registered at the impact velocity of 370 m/s (Fig. 5.2) corresponds to the equilibrium regime of dynamic deformation when the velocity defected at the plateau of the compressive pulse is comparable to the velocity variation. In order to verify the criterion (5.36), we should use the steepest part of the velocity variance profile AB where the rate of change $\mathrm{d}D/\mathrm{d}t$ is maximised. The mean velocity profile of this time interval corresponds to the part A′B′. For parts AB and A′B′, one obtains:

$D = 36.5$ m/s, $u = 250$ m/s, $\mathrm{d}D/\mathrm{d}t = 0.5 \times 10^{11}$ cm/s^2, $\mathrm{d}u/\mathrm{d}t = 2 \times 10^{11}$ cm/s^2. Substitution into Eq. (5.36) yields: $(D/u)\left(\frac{\dot{D}}{\dot{u}}\right) < 1$; that is, the criterion for structural transition is not fulfilled.

Let us perform the analogous calculations for the particle velocity variation and mean velocity profiles registered at an impact velocity of 451 m/s (Fig. 5.3). For the parts AB and A′B′ that correspond to the maximum rate of change of the velocity variance and mean particle velocity, one obtains:

$D = 46.2 \times 10^2$ cm/s, $u = 295 \times 10^{11}$ cm/s, $\mathrm{d}D/\mathrm{d}t = 0.87 \times 10^{11}$ cm/s^2, $\mathrm{d}u/\mathrm{d}t = 0.1 \times 10^{11}$ cm/s^2. Substituting these values into Eq. (5.36) yields: $(D/u)/\left(\frac{\dot{D}}{\dot{u}}\right) > 1$, that is, the criterion for structural transition is fulfilled. In Fig. 5.3, the moment

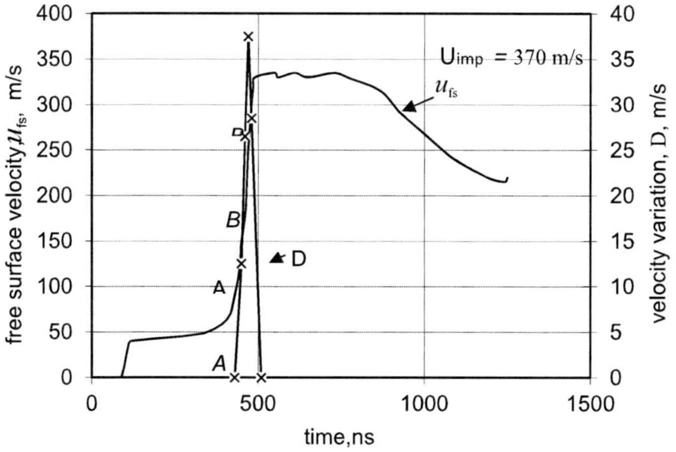

Fig. 5.2 Free surface velocity profiles for the mean velocity and the velocity variance for D16 Al alloy target loaded at the impact velocity of 370 m/s

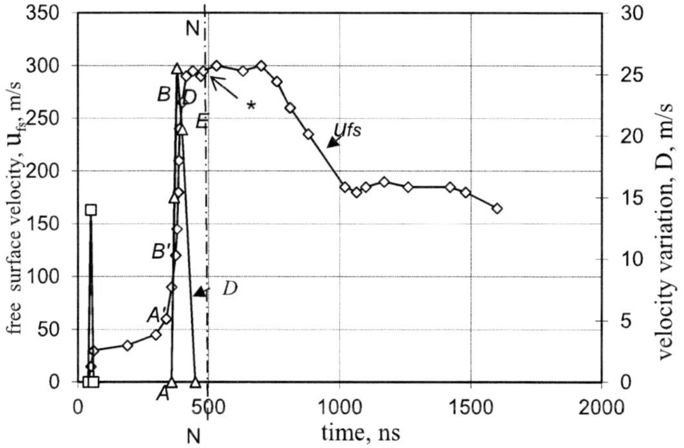

Fig. 5.3 Free surface velocity profiles of mean velocity and velocity variance for D16 Al alloy target loaded at the impact velocity of 451 m/s

of structural transition is indicated by the symbol "*". The incubation time for that transition equals ~50 ns.

5.4 Resonance Interaction of Structures and Shock Waves

Mathematically, the processes of the interaction of external noise and a nonlinear system are described with the well-known evolutional model [4]:

$$\dot{X}(t) = \lambda(t)X(t) - X^2(t) \tag{5.37}$$

where the parameter $\lambda(t)$ characterises the influence of external conditions. It consists of two parts, one of which, λ_0, is regular, while another, $\xi(t)$, is random:

$$\lambda = \lambda_0 + D\xi(t) \tag{5.38}$$

The propagation of a shock wave in a heterogeneous medium is a typical example of the interaction of a nonlinear system and external noise. Specifically, the dislocation structure can also be considered as a heterogeneous medium. Orowan's equation determines the plastic deformation for such a medium:

$$\varepsilon = B\rho l \tag{5.39}$$

Here, ρ is the density of mobile dislocation, l is their mean run, and B is the Burgers vector. Differentiation with respect to time yields

$$\mathrm{d}\varepsilon/\mathrm{d}t = B\,l(\mathrm{d}\rho/\mathrm{d}t) + B\,\rho(\mathrm{d}/\mathrm{d}t). \tag{5.40}$$

In accordance with dislocation dynamics, the rate of change of the dislocation density, $\mathrm{d}\rho/\mathrm{d}t$, is determined by the equation [5]:

$$d\rho/\mathrm{d}t = \alpha\rho - \beta\rho^2 \tag{5.41}$$

The parameter α characterises the rate of nucleation of dislocations, and β is a coefficient that considers the locking of dislocations. In the case of uniaxial strain, the deformation can be written in the form:

$$\varepsilon = u/C_p, \tag{5.42}$$

where u is the particle velocity and C_p is the shock wave velocity. For the steady plastic wave, C_p is a constant, and the strain rate equals:

$$\frac{\mathrm{d}\varepsilon}{\mathrm{d}t} = \frac{1}{C_p}\frac{\mathrm{d}u}{\mathrm{d}t}. \tag{5.43}$$

From (5.41) to (5.43), one obtains:

$$\frac{\mathrm{d}u}{\mathrm{d}t} = C_p B\frac{\mathrm{d}l}{\mathrm{d}t} + \alpha u - \left(u^2\right)\beta/BlC_p. \tag{5.44}$$

If the dislocation run is constant, Eq. (5.44) takes the form of the evolutional model (5.37)

$$\frac{du}{dt} = \alpha u - \chi u^2, \tag{5.45}$$

where $\chi = \beta/BlC_p$. To account for an interaction of random character between a shock wave and a nonlinear medium, parameter α may be presented in the form of a sum, by analogy with (5.38):

$$\alpha(t) = \alpha + D\xi(t), \tag{5.46}$$

where $\xi(t)$ is a random component and D characterises its intensity. Incorporating this random component into Eq. (5.45) transforms that equation into a stochastic differential equation [4]:

$$du = (\alpha u - \chi u^2)dt + Du \cdot dW, \tag{5.47}$$

The solution to this equation can be found by converting it to the Fokker–Planck equation [6]:

$$\frac{d}{dt} f(u, t) = -\frac{\partial}{\partial u})[F_1 f(u, t)] + \frac{D^2}{2}\frac{\partial^2}{\partial u^2}[F_2 f(u, t])], \tag{5.48}$$

where F_1 and F_2 are the diffusive coefficients of the Fokker–Planck equation:

$$F_1 = \alpha u - \chi u^2 + \frac{D^2}{2}u; \; F_2 = D^2 u^2. \tag{5.49}$$

The extremes of the probability density define the position of a phase transition, which can be found from the equation:

$$\alpha u - \chi u^2 - (D^2/2)u = 0. \tag{5.50}$$

This equation has two roots:

$$(1) \;\; u_{m1} = 0; \;\; (2). \;\; u_{m2} = \frac{1}{\chi}\left(\alpha - \frac{D^2}{2}\right). \tag{5.51}$$

The first root indicates that the probability density has an extreme at a particle velocity of zero. The second root includes two conditions concerning the behaviour of the system under consideration. When the noise is absent, that is, the particle velocity variance equals zero, the resonance condition $\alpha = \chi u$ is met. This condition supposes that the spatial scale of the macroscopic evolution of the system, $u/\alpha = \tau_f u$, coincides with the mean run of dislocations

$$l_{\mathrm{ms}} = \tau_f u. \tag{5.52}$$

In accordance with the physical nature of the mesoscale processes, at high deformation or high strain rate, a space polarisation of the dislocation ensemble occurs. Under action of load stress (see Chap. 4) fields, the initially uniform and quasi-neutral in summary sign distribution of dislocations rebuilds with the quasi-periodical separation of positive and negative dislocations. In the case of the shock wave propagation, the condition $\alpha = \chi u$ constrains the width of the plastic front to be equal to the period of the polarised mesostructures, which is a typical resonance situation. In this case, the first diffusion coefficient (5.49) equals

$$F_1 = \frac{D^2}{2} u. \tag{5.53}$$

In contrast, by definition, the first diffusion coefficient of the Fokker–Planck equation equals [5.1)]:

$$F_1 = \frac{\mathrm{d}\langle \Delta u \rangle}{\mathrm{d}t} \tag{5.54}$$

where Δu is the change of particle velocity over the time interval Δt, which is much longer than the duration of individual interactions between particles but much shorter than the time for macroscopic evolution of the system (in the case considered here, the rise time of the shock front). If $\Delta u = \mathrm{D}$, Eq. (5.54) takes the form:

$$F_1 = \frac{\mathrm{d}D}{\mathrm{d}t}. \tag{5.55}$$

Using (5.53) yields

$$\frac{1}{u}\frac{\mathrm{d}D}{\mathrm{d}t} = \frac{D^2}{2}, \tag{5.56}$$

with the condition of

$$\frac{\mathrm{d}D}{\mathrm{d}u} = \frac{\frac{\mathrm{d}D}{\mathrm{d}t}}{\frac{\mathrm{d}u}{\mathrm{d}t}} \gg \frac{D}{u}. \tag{5.57}$$

The left-hand side of this equation may be presented as follows:

$$\frac{1}{u}\frac{\mathrm{d}D}{\mathrm{d}t} = \frac{\mathrm{d}}{\mathrm{d}t}\left(\frac{D}{u}\right). \tag{5.58}$$

Indeed, for the high-velocity processes, $\frac{\dot{D}}{\dot{u}} >> D/u$, so one obtains:

$$\frac{\mathrm{d}}{\mathrm{d}t}\left(\frac{D}{u}\right)=\frac{1}{u}\frac{\mathrm{d}D}{\mathrm{d}t}+\frac{\mathrm{d}u}{\mathrm{d}t}D=\frac{1}{u}\frac{\mathrm{d}u}{\mathrm{d}t}\left(\frac{\frac{\mathrm{d}D}{\mathrm{d}t}}{\frac{\mathrm{d}u}{\mathrm{d}t}}+\frac{D}{u}\right)$$
$$=\frac{1}{u}\frac{\mathrm{d}u}{\mathrm{d}t}\left(\frac{\mathrm{d}D}{\mathrm{d}u}+\frac{D}{u}\right)\simeq\frac{1}{u}\frac{\mathrm{d}D}{\mathrm{d}t} \tag{5.59}$$

Then, the condition for the extremes of the probability density takes the form:

$$\alpha=\mathrm{d}(D/u)/\mathrm{d}t. \tag{5.60}$$

The left-hand side of this equation is the average macroscopic strain rate at the plastic front $\frac{\mathrm{d}\varepsilon_{\mathrm{mc}}}{\mathrm{d}t}=\frac{1}{\tau_f}$. Thus, when the resonance condition $l_{\mathrm{ms}}=\tau_f u$ is met, the rate of change of the variation coefficient equals the macroscopic strain rate.

5.5 Discussion

Estimates of the spatial scale for 3D structures based on analysis of the free surface velocity profiles for different impact velocities are presented below. We calculate the spatial period of polarised dislocation structure, accepting that the polarisation happens at the elastic precursor. The velocity of dislocations can be taken as equal to the transverse sound speed in aluminium: $\mathrm{V_d}=3.2\times10^5$ cm/s. The average rise time of the elastic precursor in the majority of tests equals $\tau_{\mathrm{el}}=7.5$ ns. Then, the spatial period of the polarised dislocation structure equals:

$$l_{\mathrm{d}}=V_{\mathrm{d}}\tau_{\mathrm{ef}}=3.2\times10^5\ \mathrm{cm/s}\times7.5\times10^{-9}\ \mathrm{s}=24\ \mu\mathrm{m}.$$

The space width of the plastic front is calculated for three velocity profiles, which were registered at the impact velocities of 265 m/s, 382 m/s and 451 m/s (Figs. 5.3 and 5.4). For the velocity profile of $U_{\mathrm{imp}}=265$ m/s (Fig. 5.6), the rise time of the plastic front is $\tau_{pl}=130$ ns, so its width is $l_{\mathrm{pl}}=\frac{1}{2}2.65\times10^4\ \mathrm{cm/s}\times1.3\times10^{-7}\mathrm{s}=:$ $17\times10^{-4}\ \mathrm{cm}=17\ \mu\mathrm{m}$.

For the velocity profile of $U_{\mathrm{imp}}=382$ m/s (Fig. 5.4b), the rise time of the plastic front equals $\tau_{n\Lambda}=125$ ns. The width of the plastic front is $l_{\mathrm{pl}}=\frac{1}{2}3.82\times10^4\ \mathrm{cm/s}\times125\times10^{-9}\ \mathrm{s}=23.8\ \mu\mathrm{m}$.

Lastly, for the velocity profile of $U_{\mathrm{imp}}=451$ m/s (Fig. 5.3), the rise time of the plastic front equals $\tau_{pl}=70$ ns. The width of the plastic front is $l_{\mathrm{pl}}=\frac{1}{2}4.51\times10^4\ \mathrm{cm/s}\times7\times10^{-8}\mathrm{s}=15.75\ \mu\mathrm{m}$.

It follows from the above estimates that the resonance condition for the polarised dislocation structure is fulfilled only for an impact velocity of 382 m/s. At the impact velocities of 265 m/s and 451 m/s, the width of the plastic front is 1.5 times smaller than the period of the mesostructure. Thus, the resonance condition for these impact velocities is not fulfilled.

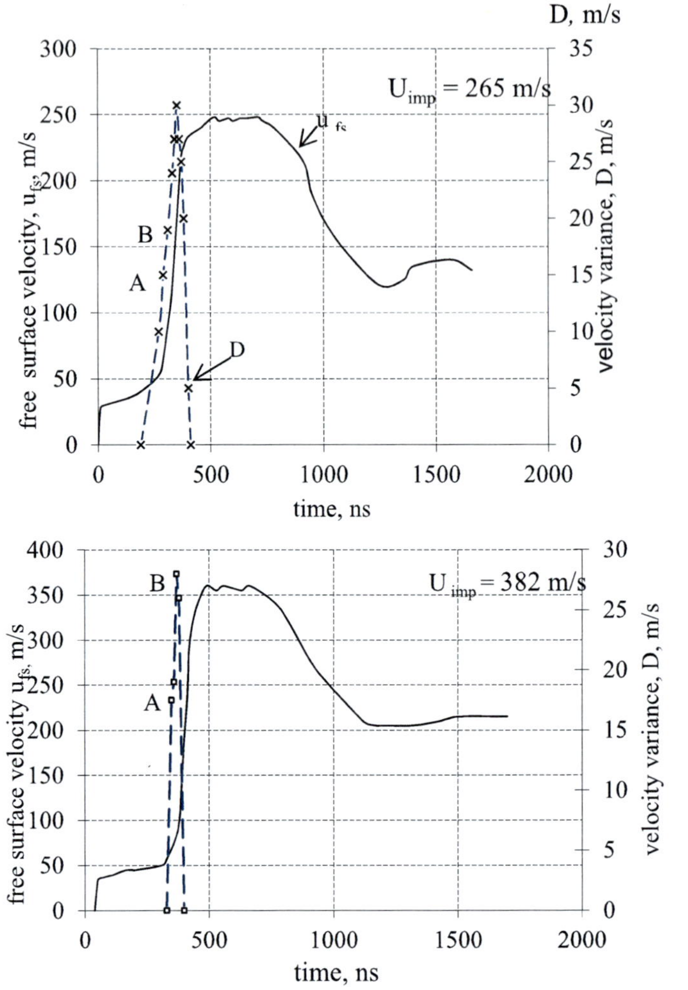

Fig. 5.4 Free surface velocity profiles of mean velocity, u_{fs}, and velocity variance, D, for a D16 Al alloy target loaded at impact velocities of 265 m/s and 382 m/s

Consider the second condition for structural transitions, which supposes the equality of the local and macroscopic strain rates. Equation (5.60) may be used to estimate the second condition for three impact velocities: 265 m/s, 382 m/s and 451 m/s (Figs. 5.3 and 5.4).

(1). $U_{imp} = 265$ m/s; at point A, $\frac{D}{u} = 0.2$; at point B, $\frac{D}{u} = 0.186$. Then, the local strain rate equals $\frac{\mathrm{d}}{\mathrm{dt}}\left(\frac{D}{u}\right) = 1.74 \times 10^6\ \mathrm{s}^{-1}$. The mean strain rate at the plastic front is $\frac{d\varepsilon_{mc}}{dt} = \frac{1}{1.3 \times 10^{-7}} = 7.7 \times 10^6\ \mathrm{s}^{-1}$, which is 4.4 times higher than the local strain rate. Thus, criterion (5.60) is not fulfilled for this impact velocity.

(2). $U_{imp} = 382$ m/s; at the point A, $\frac{D}{u} = 0.22$; at the point B, $\frac{D}{u} = 0.34$; the local strain rate equals $\frac{\mathrm{d}}{\mathrm{dt}}\left(\frac{D}{u}\right) = 0.8 \times 10^7\ \mathrm{s}^{-1}$. The mean strain rate at the plastic front is $\frac{\mathrm{d}\varepsilon_{mc}}{\mathrm{d}t} = \frac{1}{1.25\times 10^{-7}} = 0.8 \times 10^7\ \mathrm{s}^{-1}$; that is, it equals the local strain rate. Thus, criterion (5.60) is fulfilled for this impact velocity.

(3). $U_{imp} = 451$ m/s; at point A, $\frac{D}{u} = 0.2\ 0.15$; at point B, $\frac{D}{u} = 0.17$; the local strain rate equals $\frac{\mathrm{d}}{\mathrm{dt}}\left(\frac{D}{u}\right) = 2.6 \times 10^6\ \mathrm{s}^{-1}$. The mean strain rate at the plastic front is $\frac{\mathrm{d}\varepsilon_{mc}}{\mathrm{d}t} = 1.33 \times 10^7\ \mathrm{s}^{-1}$, that is, approximately twenty times higher than the local strain rate. Thus, criterion (5.60) is not fulfilled for this impact velocity.

The above estimates show that both resonance conditions are only accomplished simultaneously at an impact velocity of 382 m/s, which corresponds to the nucleation of 3D structures (see Chap. 12).

5.6 Conclusions

Shock tests of D16 aluminium alloy reveal the following features of dynamic response to shock loading:

1. The region of strain rates corresponding to impact velocities of 85-451 m/s is subdivided into two sub-regions at the impact velocity of 382 m/s. In the first sub-region (85-382 m/s), the velocity defect remains constant at ~20 m/s. In the second sub-region, corresponding to within impact velocity regon of 382-451 m/s, the velocity defect grows with increasing impact velocity. This region corresponds to the irreversible regime of interscale momentum and energy exchange.
2. At the impact velocity of 382 m/s, the velocity variance equals the velocity defects, corresponding to the equality of the local strain rate and the macroscopic strain rate, just as that impact velocity corresponds to maximum spall strength of D16 Al alloy.
3. The theoretical and experimental results show that the criterion (5.36) includes both the ratio D/u and the ratio $\frac{\mathrm{d}D/\mathrm{d}t}{\mathrm{d}u/\mathrm{d}t}$, and the second ratio plays the main role. In contrast, the first ratio is the weight coefficient. This means that in the dynamic processes, the rate of change of the particle velocity dispersion is the basic control parameter of the shock wave process.

References

1. Hubburd J (1961) Proc Roy Soc London A 260:114–126
2. Kihara T, Aono O (1963) J Phys Soc Japan 18:837–851
3. Khantuleva TA, Meshcheryakov YuI (2016) On the instability of plastic flow at the mesoscale during high-velocity deformation of solid. Phys Mesomechanics 19(4):5–13 (Russian journal)

4. Horsthemke W, Lefever R (1984) Noise-induced transitions. Springer, NY, 480p
5. Gilman JJ. Microdynamis of plastic fow at constant stress. J Appl Phys 33(9):2772–2776 (062)
6. Chandrasekar S (1947) Stochastic problems in physics and astronomy. M. IL, 168p

Chapter 6
Multiscale Modelling of Steady Shock Wave Propagation

Abstract In this chapter, a combination of deterministic and statistical approaches is used for the description of shock wave propagation in solids. The shock wave model of Duval–Taylor incorporates both the dislocation mechanism of stress relaxation and mesoscopic mechanism. This provides much better coincidence of experimental velocity profile and theoretical profile.

6.1 Introduction

The problem of the adequate description of dynamic plasticity through ECD is currently under investigation. A critical issue is that direct transition from the scale of atomic dislocations to the macroscale is impossible [1, 2]. As an additional mechanism of deformation, the concept of the mesoscale is suggested in the quasistatic region of strain rates [2]. The carriers of deformation, mesoparticles, are accepted as the structures that occupy an intermediate position between the atom-dislocation scale and the macroscale. As for shock wave processes, mesomechanics cannot describe the total spectrum of shock-induced deformation mechanisms at the intermediate scale. Thus, the first problem is describing the transition from one scale to another. Hard coupling between dynamic variables at the macroscale and the atom-dislocation scale does not work. Specifically, the well-known formula [3]

$$\Delta\rho^o \rho_{ik}^{+} - \rho_{ik}^{-} = -\frac{1}{b} e_{ijl} \frac{d\varepsilon_{jk}}{dx_l} \tag{6.1}$$

which links the crystal curvature (macroscopic scale) and the density of dislocations (microscale) can be used only for perfect crystals. Here, ρ_{ik} and ρ_{ik} are the densities of positive and negative dislocations, b is the value of the Burgers vector, and ε_{jk} is the component of deformation.

A combination of deterministic and statistical approaches is thought to be an adequate description of high strain rate processes. Instead of hard coupling in the form of (6.1), the transition must incorporate statistical variables which could provide a flexible linkage between neighbouring scales. Also, there must be test regimes of

Y. Meshcheryakov, *Multiscale Mechanics of Shock Wave Processes*, Shock Wave and High Pressure Phenomena, https://doi.org/10.1007/978-981-16-4530-3_6

dynamic deformation where the theoretical approach can be verified. In this relationship, one of the most convenient tests is known to be a steady shock wave, which is often used to verify constitutive equations, determine the effective viscosity and develop computational codes [4–7].

6.2 Coupling Between the Strain Rate and the Mesoparticle Velocity Distribution

In Fig. 6.1, the time-resolved free surface velocity profile for the plastic front, $u_{fs}(t)$, the velocity variation profile, $D(t)$, and the strain rate profile, $\frac{d\varepsilon(t)}{dt}$, are plotted.

It follows from Fig. 6.1 that

$$D = R\frac{d\varepsilon(t)}{dt}, \tag{6.2}$$

where R is a proportionality coefficient. In our approach, the relationship (6.2) should be considered as a hypothesis. The correctness of this hypothesis can be verified after comparing the calculated and experimental velocity profiles.

By analogy with the temperature that results from the random motions of atoms, the particle velocity dispersion is a quantitative characteristic of the intensity of random motions of mesoparticles (in the Western literature, the mesoparticle velocity dispersion is referred to as the "granular temperature"). Even so, the particle velocity

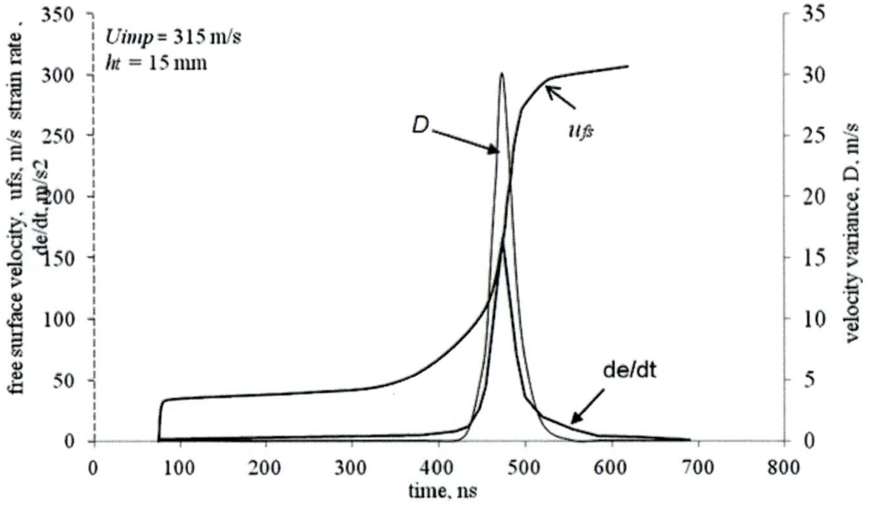

Fig. 6.1 Free surface velocity profile, u_{fs}, velocity variation profile, D, and strain rate profile, $d\varepsilon/dt$, for 15 mm target of D16 aluminium alloy at the impact velocity 315 m/s

dispersion provides additional contributions to the spherical components of the pressure and the deformation. The mean deformation in the shock wave and the contribution to the deformation owing to the velocity variance can be expressed in the following form:

$$\varepsilon = \frac{u}{C_0}; \ \Delta\varepsilon^{(1)} = \frac{D}{C_0}, \tag{6.3}$$

Substituting Eq. (6.2) into (6.3) yields

$$\Delta\varepsilon^{(1)} = \frac{R}{C_0}\frac{\mathrm{d}\varepsilon(t)}{\mathrm{d}t}. \tag{6.4}$$

One possible theoretical basis for the relationship in (6.2) may be the hypothesis based on the supposition that a crystalline structure consists of two sub-lattices that have similar characteristics and are coupled by nonlinear forces of internal interaction Q. It is natural to assume that Q should be defined through a nonlinear coupling between the components and the dissipative terms:

$$Q = Q_1(\nu_1 - \nu_2) + Q_2(\dot{\nu}_1 - \dot{\nu}_2). \tag{6.5}$$

The model form of the graph of $Q_1(\nu_1 - \nu_2)$ is depicted in Fig. 6.2.

The curve in Fig. 6.2 has a nontrivial ($w = \nu_1 - \nu_2^1 0$) equilibrium position, indicating the possibility for a new stable state of the material. The interaction force

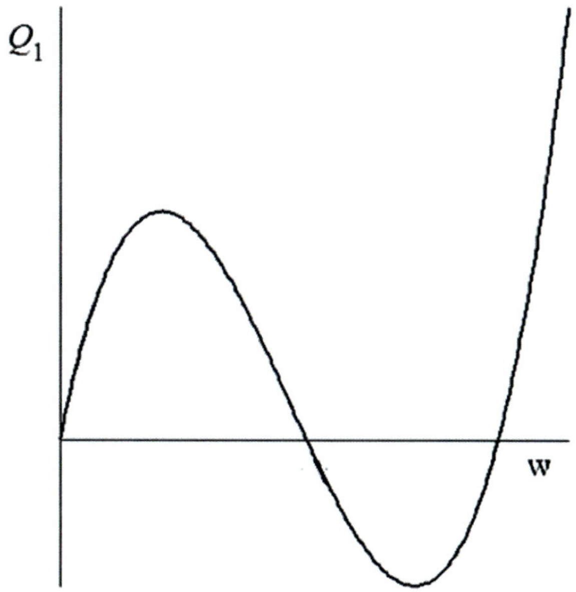

Fig. 6.2 Nonlinear interaction forces

depends on the properties of the two-component medium. Suppose the change of the structure of the medium is determined by the displacements of these sub-lattices relative to each other, w. When w achieves the critical value, the nonlinear force transfers to the unstable branch, which results in the complex dynamic behaviour of the system. Considering both the nonlinearity of the interaction and the periodicity of the internal structure, the simplest form of the interaction force can be expressed as follows

$$Q(w, \dot{w}) = K \sin \lambda w + n\dot{w},$$

where K is a maximum value of Q, the coefficient n characterises the dissipation, and λ is a period in the crystalline lattice. In essence, the function w (relative displacement) appears to be a latent degree of freedom, which may have a serious influence on the dynamics of the system. It plays the same role as dispersion, which is why $\frac{\mathrm{d}w}{\mathrm{d}t}$ is considered the analogue of the particle velocity variation D. It is convenient to rewrite Eq. (6.5) with respect to the functions $v = \frac{\rho_1 v_1 + \rho_2 v_2}{\rho_1 + \rho_2}$ and $w = v_1 - v_2$. In the case of small deformations, it can be accepted that for each sub-lattice, Hook's law is in effect. Then, in the one-dimensional case, the balanced equations for the momentum of each component are as follows:

$$E_1 \frac{\mathrm{d}^2 v_1}{\mathrm{d}x^2} - Q = \rho_1 \frac{\mathrm{d}^2 v_1}{\mathrm{d}t^2};$$
$$E_2 \frac{\mathrm{d}^2 v_2}{\mathrm{d}x} + Q = \rho_2 \frac{\mathrm{d}^2 v_2}{\mathrm{d}t^2}.$$

Here, E_i is the Young's modulus for the i-lattice, ρ_i is the density of the components in the initial state, and v_i is the displacement. The velocity of the centre of mass for the system is defined as

$$u = \frac{\rho_1 \frac{\mathrm{d}v_1}{\mathrm{d}t} + \rho_2 \frac{\mathrm{d}v_2}{\mathrm{d}t}}{\rho_1 + \rho_2}. \tag{6.6}$$

Substitution into (6.5) yields

$$\frac{\mathrm{d}^2 w}{\mathrm{d}x^2} - \frac{\rho_1 \rho_2}{\rho_1 + \rho_2} \left[\frac{1}{E_1} + \frac{1}{E_2} \right] \frac{\mathrm{d}^2 w}{\mathrm{d}t^2} = \left[\frac{\rho_1}{E_1} - \frac{\rho_2}{E_2} \right] \frac{\mathrm{d}^2 v}{\mathrm{d}t^2} + \left[\frac{1}{E_2} + \frac{1}{E_1} \right] Q, \tag{6.7}$$

where

$$w = v_1 - v_2, \ \frac{\mathrm{d}v}{\mathrm{d}t} = u, \ v_1 = v + \frac{\rho_2}{\rho_1 + \rho_2} w, \ v_2 = v - \frac{\rho_1}{\rho_1 + \rho_2} w, \tag{6.8}$$

Specifically, at the wave front, the relative displacement satisfies the d'Alembert operator. It follows from (6.7) that

$$\frac{\mathrm{d}^2 w}{\mathrm{d}x^2} - \frac{\rho_1 \rho_2}{\rho_1 + \rho_2}\left[\frac{1}{E_1} + \frac{1}{E_2}\right]\frac{\mathrm{d}^2 w}{\mathrm{d}t^2} = 0,$$
$$\left[\frac{\rho_1}{E_1} - \frac{\rho_2}{E_2}\right]\frac{\mathrm{d}^2 v}{\mathrm{d}t^2} + \left[\frac{1}{E_2} + \frac{1}{E_1}\right]Q = 0 \tag{6.9}$$

or

$$\frac{\mathrm{d}^2 v}{\mathrm{d}t^2} = \frac{E_1 + E_2}{E_1\rho_2 - E_2\rho_1} Q\left(w, \frac{\mathrm{d}w}{\mathrm{d}t}\right). \tag{6.10}$$

In the linear approximation for the interaction force in the unstable branch of (6.9), one obtains

$$C_0 \frac{\mathrm{d}\varepsilon}{\mathrm{d}t} = \frac{E_1 + E_2}{E_1\rho_2 - E_2\rho_1} Q\left(w, \frac{\mathrm{d}w}{\mathrm{d}t}\right), \tag{6.11}$$

or

$$\frac{\mathrm{d}\varepsilon}{\mathrm{d}t} = \frac{1}{\rho_1}\frac{\frac{\rho_1}{\rho_2}C_1^2 + C_2^2}{C_1^2 - C_2^2}\frac{Q\left(w, \frac{\mathrm{d}w}{\mathrm{d}t}\right)}{C_0}. \tag{6.12}$$

A transition from a two-component to a multi-component medium yields

$$\left\langle \frac{\mathrm{d}\varepsilon}{\mathrm{d}t} \right\rangle = R\frac{D}{C_0}, \tag{6.13}$$

where $\rho_1 c_1^2 = E_1$, $\rho_2 c_2^2 = E_2$, and $R = \left(K\frac{E_1+E_2}{E_1\rho_2 - E_2\rho_1}\right)^{-1}$ is the proportionality coefficient, which depends on the parameters of the medium. Equation (6.13), which links the particle velocity variance, *D*, and the mean macroscopic strain rate, coincides with the empirical relationship, Eq. (6.2) (see Fig. 6.1). Thus, the additional contribution to the summary plastic strain rate due to the velocity variation is determined by Eq. (6.4).

6.3 The Relaxation Model for a Steady Shock Wave

This section presents analytical solution to the plastic steady shock wave based on dislocation dynamics. The constitutive equation for one-dimensional shock wave propagation has been introduced by Taylor [8] and Duvall [9] in the form:

$$\sigma_t - \rho C_l^2 \varepsilon_t = -F. \tag{6.14}$$

Here, F is a relaxation function, and the index t indicates differentiation with respect to time.

$$F = \frac{8}{3} b N_d V_d, \tag{6.15}$$

and γ_t^p is the shear plastic strain rate. The latter is expressed in terms of the density of mobile dislocations, N_m, and the dislocation velocity, V_d, through Orowan's equation:

$$\gamma_t^p = b N_m V_d, \tag{6.16}$$

where b is the Burgers vector. The following law for dislocation density is accepted [10, 11]:

$$N_m = N_{m0} + \alpha \gamma^p, \tag{6.17}$$

where N_{m0} is the initial density of mobile dislocations, and α is the coefficient of dislocation multiplication.

There is experimental evidence for a linear relationship between the dislocation velocity and the shear stress over a wide range of strain rates [11, 12]:

$$V_d = \frac{b(\tau - \tau_0)}{B}, \tag{6.18}$$

where B is the drag coefficient for dislocation motion, τ is the applied shear stress, and τ_0 is the back stress. The constitutive Eq. (6.19) then takes the form:

$$\sigma_t - \rho C_l^2 \varepsilon_t = \gamma_* \mu (1 - M\gamma^p)(\tau - \tau_0), \tag{6.19}$$

where

$$\gamma_* = \frac{b^2 N_{m0}}{B}; \; M = \frac{\alpha}{N_{m0}}, \; \tau_* = \tau_0/\mu. \tag{6.20}$$

In terms of total and shear plastic deformation, Eq. (6.20) can be presented in the form:

$$\gamma_t^p = \gamma_* \mu (1 + M\gamma_p)\left[(\varepsilon - 2\gamma^p) - \tau_*\right]. \tag{6.21}$$

Together with the momentum and mass balance equations

$$\rho u_t + \sigma_x = 0, \tag{6.22}$$

$$u_x + \varepsilon_t = 0, \tag{6.23}$$

equations (6.22)–(6.24) form a system of equations for the one-dimensional dynamic deformation of isotropic material. For steady wave propagation with constant velocity C_0, one independent variable can be introduced: $z = x - C_0 t$. In this case, the system takes the form:

$$\begin{aligned} &\rho C_0 u_z - \sigma_z = 0, \\ &C_0 \varepsilon_z - u_z = 0, \\ &C_0 \gamma_z^p = \gamma_* \mu (1 + M\gamma^p)\big[(\varepsilon - 2\gamma^p) - \tau_*\big] \end{aligned} \tag{6.24}$$

Equation (6.25) can be reduced to the second-order nonlinear differential equation for particle velocity u:

$$u_z + L_1 u + L_2 u^2 + L_3 = 0, \tag{6.25}$$

where

$$\begin{aligned} L_1 &= \frac{1}{C_0}\gamma_* \mu (a_1 - 2 - M\tau_*),\, L_2 = \frac{\gamma_* \mu M}{a_1 C_0^2}(a_1 v2),\, L_3 = a_1 \gamma_* \mu \tau_*, \\ a_1 &= \frac{8\mu}{3}\frac{1}{\rho(C_l^2 - C_0^2)} \end{aligned} \tag{6.26}$$

The solution to this equation is well-known [13]:

$$u = U_0 \left\{ 1 + \exp\left[\frac{U_0 \gamma_* M (a_1 - 2)}{a_1 C_0} z \right] \right\}^{-1} \tag{6.27}$$

where U_0 is the value of the particle velocity at $z = -\infty$. Equation (6.25) describes the propagation of a steady plastic front from the region $z = -\infty$ to the region $z = +\infty$. The steepness of the shock front is determined by the parameters of the dislocation structure, which are incorporated in the form of the initial density of the mobile dislocation, N_{m0}; the coefficient of multiplication of dislocations, α; and the drag coefficient, B.

To compare the solution to (6.25) with experiments, we use the steady wave velocity profile for a 15 mm D16 aluminium alloy target loaded at the impact velocity of 315 m/s. The data for the D16 Al alloy and the mechanical parameters of shock loading are provided in Tables 6.2, 6.3 and 6.4. The parameters for the dislocation structure are taken from [4, 11, 12].

The calculated velocity profiles for two values of the initial dislocation density are presented together with the experimental profile in Fig. 6.3. The experimental profile does not coincide with the calculated profiles for the initial densities of mobile dislocations of 10^8 and 10^9 cm^{-2}. In reality, the initial values of mobile dislocation density are two to three orders of magnitude smaller, so dislocation mechanisms alone cannot adequately describe the plastic deformation exhibited in shock loading experiments.

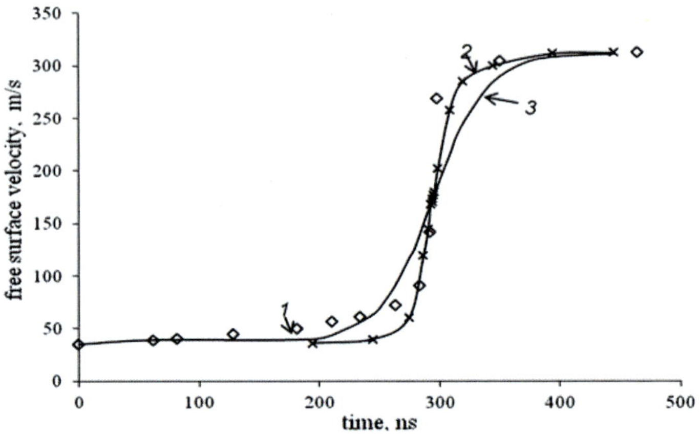

Fig. 6.3 Free surface velocity profiles calculated with the Duval–Taylor model for the initial dislocation densities of $N_{m0} = 10^8\ \text{cm}^{-2}(1)$ and $N_{m0} = 10^9\ \text{cm}^{-2}(2)$.

The above model concerns the steady wave state, so one should verify whether the particle velocity profile is the steady wave. We will calculate the so-called Bland's number, $B' = h_t/\delta$, which is the ratio of the sample thickness h_t to the steady wave thickness, δ [6]. The latter determines the distance necessary for the shock wave to become steady. The time for the plastic front to become steady, t_0, has been found in [14], yielding

$$\frac{t_0}{\Delta} = \frac{4\frac{C_h}{C_0}\left(1 - \frac{C_h^2}{C_0^2}\right)^{-1}}{M\left(g\rho C_0 U_{\text{imp}}\mu - \frac{\tau_0}{\mu}\right)\ln 19}. \tag{6.28}$$

Here, $g = \frac{1-2\nu}{2(1-\nu)}$, ν is the Poisson ratio, C_h is the bulk velocity, and Δ is the rise time of the steady wave front. For the shock loading of a 15 mm target loaded at the impact velocity of 315 m/s, the rise time of the particle velocity profile, $\Delta = 125$ ns, then $t_0 = 0.69 \times 10^{-6}$ s. With the above model parameters, the steady wave thickness is $\delta = C_0 t_0 = 3.8 \times 10^{-1}$ cm, and Bland's number equals $B' = \frac{h_t}{\delta} = \frac{1.5}{3.8\times10^{-1}} = 3.95 > 1$. Hence, this velocity profile is a steady wave.

6.4 Account for the Mesoscopic Effects

Shock wave experiments [6, 7, 15, 16] and theoretical investigations [17, 18] show that the propagation of a shock front in a heterogeneous medium is non-uniform in the velocity space process. The question is how to consider the non-uniformity in the macroscopic response of the solid to shock loading. This question directly relates to the problem of a multiscale description of dynamic deformation, which is certainly a stochastic process and must include stochastic variables. In [17], the statistical features of a shock-deformed medium are taken into account by incorporating the normal law for yield limit. The two statistical moments of the distribution function—mean yield limit and variation of yield limit—are used for describing shock wave propagation under reloading and unloading. Unfortunately, the mean-square deviation of the yield limit cannot be measured during the process of dynamic straining. Therefore, these values can be used only as fitting parameters. Modelling of shock wave propagation with cellular automata [19] shows that the time to reach the equilibrium state where the particle velocity is established in copper equals 11.5 ns, an order of magnitude smaller than the duration of the wave front. In this situation, the PVDF can be considered to be at equilibrium within the plastic front. It can be characterised by two statistical moments: the mean particle velocity and the particle velocity dispersion, the first and the second statistical moments of the PVDF. Both values can be measured in real time. In steady shock waves, the maximum value of the mean velocity variation coincides with the middle of the plastic front [20]. Physically, this means that the intensity of the velocity pulsations is proportional to the mean particle acceleration.

Our next step is to use Eq. (6.4) as an additional relaxation term in the constitutive Eq. (6.19). The right-hand side of this equation includes two terms. The first term, $\Delta\varepsilon_t^d$, describes the contribution of dislocations into the summary strain rate (an additive mechanism), whereas the second term, $\Delta\varepsilon_t^{(1)}$, is the collective mechanism due to particle velocity dispersion:

$$\sigma_t - \rho C_l^2 \varepsilon_t = -2\mu(\varepsilon_t^d + \Delta\varepsilon_t^1) \tag{6.29}$$

Substituting (6.8) into (6.34) yields

$$\sigma_t - \rho C_l^2 \varepsilon_t = -2\mu\left(\varepsilon_t^d + \frac{R}{C_0}\varepsilon_{tt}\right). \tag{6.30}$$

Together with the balance Eqs. (6.23)–(6.24), the constitutive Eq. (6.30) forms a closed equation system that describes the propagation of a steady plastic front in a heterogeneous medium, taking into account the mesoscopic effects. After the transition to a single independent variable, one obtains

$$\begin{aligned} & u_{zz} - \left[\frac{2\gamma^*\mu}{c_0} - \frac{1}{ar}\right]u_z - \frac{\gamma^*\mu M}{ac_0^2}[a-4]uu_z \\ & + \frac{2\gamma^*\mu M r}{c_0^2}u_z^2 + \frac{\gamma^*\mu}{ac_0 r}[a-2]u - \frac{\gamma^*\mu M}{a^2 c_0^2 r}[a-2]u^2 = 0, \\ & a = \frac{8\mu}{3\rho\left(c_l^2 - c_0^2\right)},\ r = \frac{3}{4}R. \end{aligned} \tag{6.31}$$

This equation can be written in the form

$$u_{zz} + L_{u_z} u_z + L_{u_z u} u u_z + L_{u_z^2} u_z^2 + L_u u + L_{u^2} u^2 = 0, \tag{6.32}$$

where

$$L_{u_z} = -\left[\frac{2\gamma^* \mu}{c_0} - \frac{1}{ar}\right] = -\frac{2\gamma^* \mu}{c_0 a}\left[a - \frac{c_0}{2\gamma^* \mu r}\right], L_{u_z u} = -\frac{\gamma^* \mu M}{a c_0^2}[a-4],$$
$$L_{u_z^2} = \frac{2\gamma^* \mu M r}{c_0^2}, L_u = \frac{\gamma^* \mu}{a c_0 r}[a-2], L_{u^2} = -\frac{\gamma^* \mu M}{a^2 c_0^2 r}[a-2]. \tag{6.33}$$

For $a = 2$, Eq. (6.33) becomes

$$u_{zz} + L_{u_z} u_z + L_{u_z u} u u_z + L_{u_z^2} u_z^2 = 0. \tag{6.34}$$

In the case of $a^1 2$, Eq. (6.32) transforms into

$$u_{zz} + \beta_{u_z} \lambda_{u_z} k_u u_z + \beta_{u_z u} \frac{\lambda_{u_z u} k_u}{u_\infty} u u_z$$
$$+ \beta_{u_z^2} \frac{\lambda_{u_z^2} k_u}{u_\infty} u_z^2 + \beta_u k_u u + \beta_{u^2} \frac{k_u}{u_\infty} u^2 = 0, \tag{6.35}$$

where

$$k_u(r) = \frac{N_0 b^2 \mu |a-2|}{a c_0 B}\frac{1}{r}, \gamma^* = \frac{N_0 b^2}{B}, M = \frac{N_0}{\alpha}, u_\infty = \frac{a c_0}{M} = \frac{a c_0 N_0}{\alpha},$$
$$\eta = \frac{c_0 B}{2 N_0 b^2 \mu r}, \beta_{u_z} = -\text{sign}(a-\eta), \beta_{u_z u} = -sign(a-4), \beta_{u_z^2} = 1,$$
$$\beta_u = \text{sign}(a-2), \beta_{u^2} = -\text{sign}(a-2), \lambda_{u_z}(r) = 2r\frac{|a-\eta|}{|a-2|},$$
$$\lambda_{u_z u}(r) = a\frac{|a-4|}{|a-2|}r, \lambda_{u_z^2}(r) = \frac{2a^2}{|a-2|}r^2. \tag{6.36}$$

The coefficients in Eq. (6.40) are presented in Table 6.1 for different parameters a and η.

Equation (6.35) can be solved numerically, but there is an analytical solution for certain values of λ_i. For $\lambda_i = 0$, Eq. (6.35) transforms into a nonlinear oscillator

$$u_{zz} + k_u u - \frac{k_u}{u_\infty} u^2 = 0, \tag{6.37}$$

A solution to this equation can be found with the following algorithm:

$$f = \cos\sqrt{k_u} z, f_z = -\sqrt{k_u} \sin\sqrt{k_u} z,$$
$$f_{zz} = -k_u f, f_z^2 = k_u \sin^2 \omega z, f^2 = \cos^2 \omega z,$$

Table 6.1 Regions of changing the paprameters, at what the analytical solutions are possible

a, η	β_{u_z}	$\beta_{u_z u}$	$\beta_{u_z^2}$	β_u	β_{u^2}	Comments
$0 < a < 2, a < \eta$	1	1	1	−1	1	There is an analytical solution
$0 < a < 2, a = \eta$	0	1	1	−1	1	There is an analytical solution
$0 < a < 2, a > \eta$	−1	1	1	−1	1	There is an analytical solution
$2 < a < 4, a < \eta$	1	1	1	1	−1	
$2 < a < 4, a = \eta$	0	1	1	1	−1	
$2 < a < 4, a > \eta$	−1	1	1	1	−1	
$a = 4, a < \eta$	1	0	1	1	−1	
$a = 4, a = \eta$	0	0	1	1	−1	
$a = 4, a > \eta$	−1	0	1	1	−1	
$a > 4, a < \eta$	1	−1	1	1	−1	There is an analytical solution
$a > 4, a = \eta$	0	−1	1	1	−1	
$a > 4, a > \eta$	−1	−1	1	1	−1	

Table 6.2 D16 aluminium alloy and its analogues

Russia, GOST 4784	USA, ASTM B209	UK, BS 1471	Germany DIN 1725
D16(1160)	2024	L.97 L98 L109 L119	AlCuMg 3.1355

Table 6.3 Mechanical characteristics for D16 aluminium alloy

Shear modulus, μ (GPa)	27
Longitudinal sound velocity, C_l (cm/s)	6.4×10^5
Quasistatic strength, σ_b (MPa)	440
Elastic proportionality limit, $\sigma_{0,2}$ (MPa)	290
Relative elongation, δ (%)	10
Dislocation drag coefficient, B (GPa s)	5×10^{-6}
Burgers vector, b (cm)	2.86×10^{-8}
Dislocation multiplication coefficient, α (cm^{-2})	2×10^{11}
Elastic precursor amplitude, $u_.$ (m/s)	37
Maximum particle velocity, u_1 (m/s)	312
Plastic front velocity, C_0 (cm/s)	5.47×10^5

$$f_z^2 + k_u f^2 = k_u,\ u = \frac{u_\infty}{f+1},\ u_z = -\frac{u_\infty f_z}{(f+1)^2},$$

$$u_{zz} = -u_\infty\left[\frac{f_{zz}(f+1)^2 - 2f_z^2(f+1)}{(f+1)^4}\right] = -u_\infty\left[\frac{f_{zz}}{(f+1)^2} - \frac{2f_z^2}{(f+1)^3}\right]$$

Table 6.4 Free surface velocity profiles calculated with the basic model, additional contribution, summary velocity profile and experimental profile for 15 mm D16 aluminium alloy target loaded at an impact velocity of 315 m/s

Time (ns)	Calculated with the basic model (m/s)	Additional contribution (m/s)	Summary profile (m\s)	Experimental profile (m/s)
400	56	−0	56	56
425	75	−15	60	61
450	102	−30	72	71.7
465	125	−15	110	108
480	173.5	0	173.5	173.5
495	235	+15	250	251
510	260	+30	290	291
545	292	+15	307	306
580	307	0	307	307

$$= -u_\infty\left[\frac{-k_u f}{(f+1)^2} - 2\frac{f_z^2 + k_u f^2 - k_u f^2}{(f+1)^3}\right] = -u_\infty\left[\frac{-k_u f}{(f+1)^2} + 2k_u\frac{f^2-1}{(f+1)^3}\right]$$

$$= -u_\infty\left[k_u\frac{1}{f+1} - \frac{3k_u}{u_\infty}\frac{u_\infty}{(f+1)^2}\right] = -k_u u + \frac{3k_u}{u_\infty}u^2. \tag{6.38}$$

For the equation

$$u_{zz} + k_u u - \frac{3k_u}{u_\infty}u^2 = 0, \tag{6.39}$$

the solution is

$$u = \frac{u_\infty}{\cos\sqrt{k_u}z + 1}, \tag{6.40}$$

When the contribution to the plastic strain rate in Eq. (6.32) is not too large, it is possible to express a simple form of the solution to the system of Eqs. (6.23)–(6.24). This system can be reduced to a second-order differential equation with square nonlinearity:

$$L_0 u_{tt} + u_t + L_1 u + L_2 u^2 + L_3 = 0 \tag{6.41}$$

where $L_0 = \frac{R}{C_0}$. For an approximate solution to this equation, one can use the fact that the addition to the basic solution (6.28) is a small value. Specifically, the maximum value of the particle velocity variance is approximately ten times smaller than the magnitude of the particle velocity (see Fig. 6.1). In this situation, the solution to Eq. (6.41) can be presented in the form

$$u = u_0 + u_1, \tag{6.42}$$

where

$$u_1 << u_0. \tag{6.43}$$

By substituting Eq. (6.42) into (6.41) and neglecting the small order terms, one obtains two equations, the first of which determines the basic solution u_0 in the form of (6.28), while the second equation determines the addition, u_1.

$$u_z^{(0)} + L_1 u^{(0)} + L_2 (u^{(0)})^2 + L_3 = 0, \tag{6.44}$$

$$u_{zz}^{(1)} + L_0 u_z^{(1)} + L_1 u^{(1)} + L_3 = 0, \tag{6.45}$$

where

$$L_0 = \frac{1}{R}, L_1 = \frac{1}{C_0 R}(a_1 - 2), L_2 = \frac{\gamma_* M}{R a_1 C_0^2}(a_1 - 2)a_1 = \frac{8\mu}{3} \frac{1}{\rho (C_l^2 - C_0^2)}. \tag{6.46}$$

The coefficient L_3 corresponds to a constant shift along the z-axis, so it may be taken as zero. The solution to Eq. (6.45) has the form

$$u_1 = A e^{-\frac{az}{2}} \sin \frac{\lambda}{2}(z - B), \tag{6.47}$$

where

$$a = L_0, \lambda^2 = a^2 - 4b, b = L_1. \tag{6.48}$$

The coefficient A in Eq. (6.47) equals the maximum value of the particle velocity variance, that is, $A = D = 30$ m/s, and the wave number in Eq. (6.47) is determined as

$$\lambda^2 = \frac{1}{R^2} - 4\frac{\gamma_*}{C_0 R}(a_1 - 2) = 6.2 \times 10^4. \tag{6.49}$$

The spatial period of particle velocity oscillations coincides with the width of the shock front. In this case, during the first half of the shock front, the additional contribution u_1 decreases the mean particle velocity of the basic solution (6.28), whereas, during the second half, the mean particle velocity is increased by the same amount.

The results of calculations are presented in Fig. 6.4. The basic solution (6.28) does not coincide with the experimental profile. The deviation of the basic solution from the experimental data is 50–60 m/s. Calculated with Eqs. (6.52)–(6.54), the profile

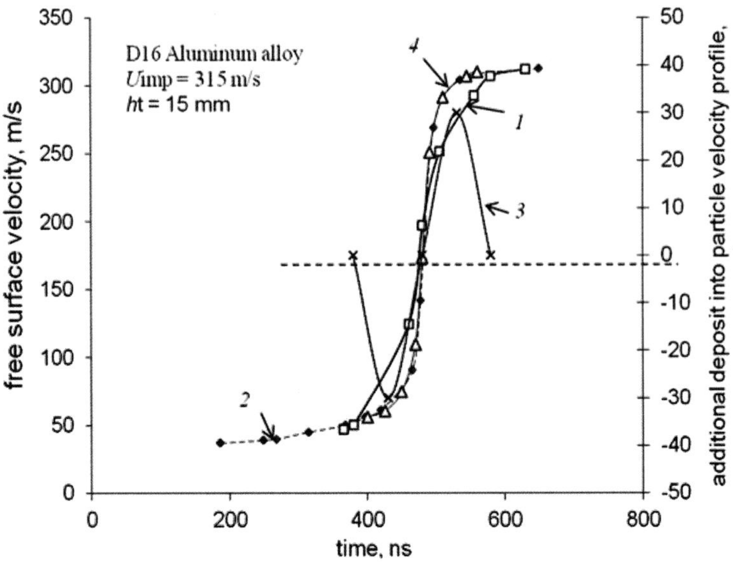

Fig. 6.4 1—Basic steady free surface velocity profile calculated with the Taylor–Duvall model; 2—experimental profile; 3—addition to basic profile; 4—resulting profile

for the same parameters of the dislocation structure coincides with the experimental profile within the overall duration of the shock front. In this case, the deviation of the calculated profile from the experimental data does not exceed 2–3 m/s, which is within the scattering of the experimental data from the velocity interferometer.

6.5 Conclusions

The additive dislocation mechanism of plastic deformation is shown to be incapable of adequately describing steady wave propagation. In order to take into account the collective mechanism, we used the hypothesis of proportionality between the particle velocity variation and the strain rate. Being incorporated into the constitutive dislocation equation of Duvall and Taylor, this contribution acts as an additional stress-relaxing mechanism. The resulting governing equation for one-dimensional propagation of the plastic front is a second-order differential equation with square nonlinearity. The solution to this equation reveals the oscillations of particle velocity over the plastic front; the period of oscillation equals the rise time of the plastic front. As a result, the steepness of the plastic front increases, so the theoretical and experimental velocity profiles for the shock front show better agreement.

References

1. Johnson JN, Jones OE, Michaels TE (1970) Dislocation dynamics and single-crystal constitutive relation: shock-wave propagation and precursor decay. J Appl Phys 41(6):2230–2239
2. Panin VE, Grinyaev YuV, Elsukova TF, Ivanchkhin AG (1982) Structural level of plastic deformation. Izvestiya Vuzov .Fizika 25(6):5–24
3. Eshby MF (1970) The deformation of plastically non-homogeneous material. Philosophical Mag 2:399–422
4. Johnson JN, Barker LM (1969) Dislocation dynamics and steady plastic wave profiles in 6061–T6 aluminum. J Appl Phys 40(11):4321–4335
5. Prieto FE, Renero C (1973) Steady shock profile in solids. J Appl Phys 44(9):4013–4017
6. Swegle JW, Grady DE (1985) Shock viscosity and the prediction of shock-wave rise times. J Appl Phys 58(2):692–701
7. Grady DE (1986) Steady rise-time and spall measurement on Uranium (3–16 GPa). In: Metallurgical applications of shock-wave and high-strain-rate phenomena (Explomet- 85). In: Murr LE, Staudhammer KR, Meyers MA (eds). Marcel Dekker, NY, pp 763–780
8. Taylor JW (1965) Dislocation dynamics and dynamic yielding. J Appl Phys 36(10):3146–3155
9. Duvall GE. Propagation of plane shock waves in a stress-relaxing medium. In: Kolsky H, Prager W (eds) Stress waves in inelastic solids. Springer, Berlin, pp 20–32
10. Gilman JJ (1968) Dislocation dynamics and the response of materials to impact. Appl Mech Rev 21(8):767–783
11. Kumar A, Hauser FE, Dorn IE (1967) Viscous drag of dislocation in aluminum at high strain rates. J Appl Phys 38(2):1863–1871
12. Johnston WG, Gilman JJ (1959) J Appl Phys 30:129–141
13. Bland DR (1965) On shock structure in a solid. J Inst Maths Appl 1:56–75
14. Gaikov AL, Meshcheryakov YuI (1989) Propagation of elastic-plastic wave in a material with dislocation kinetics of plastic deformation. J Appl Mech Tech Phys 6:158–162 (Russian edition)
15. Asay JR, Barker LM (1974) Interferometric measurements of shock-induced internal particle velocity and spatial variation of particle velocity. J Appl Phys 1974(45):2540–2546
16. Mescheryakov YuI, Divakov AK (1994) Kinetics of mesostructure and strain-rate dependence of materials. Dymat J 1(4):271–287
17. Lipkin J, Asay JR (1977) Reshock and release of shock-compressed 6061–T6 aluminum. J Appl Phys 48(1):182–189
18. Asay JR, Chhabildas LC (1989) Determination of the shear strength of shock compressed 6061-T6 aluminum. In: Murr LE, Staudhammer KS, Meyers MA (eds) Shock waves and high-strain rate phenomena. Marcel & Dekker, NY, pp 763–780
19. Yano K, Horie Y-Y (1999) Discrete element modeling of shock compression of polycrystalline copper. Phys Rev B 59(21):13672–13680
20. Meshcheryakov YuI, Divakov AK, Zhigacheva NI, Makarevich IP, Barakhtin BK (2008) Dynamic structures in shock-loaded copper. Phys Rev B 78:64301–64316

Chapter 7
On the Chaotic and Translational Motions of Elementary Carriers of Deformation at the Mesoscale

Check for updates

Abstract This problem concerns the propagation of a plane shock wave in a relaxing medium. We consider the situation when only the mesoscopic mechanism of stress relaxation is taken into account. We accept that the relaxation of the equilibrium state of the heterogeneous medium flows due to the meso–macromomentum exchange, the quantitative characteristic of which is the velocity defect a theoretical investigation of the oscillation regime at the mesoscale is carried out. As distinct from the Duvall–Taylor approach and mixed approach, in the present investigation, the stress relaxation in a dynamically deformed medium is realised only through the motion of ECD at the mesoscale. Solution to governed equations results in oscillatory regime of shock wave propagation. Where observed, the oscillations are excited at the top of the plastic front. The oscillations are initiated at the impact velocity, which is higher than the threshold for the structurally unstable transition, $U_{ins.}$ Although the oscillations are initiated due to the interaction of stochastic features of the dynamically deformed structure in the form of the particle velocity dispersion, the meso–macrotransition itself happens due to the swinging high-frequency oscillations of the mesostructure. Thus, it may be concluded that the meso- to macrostructural transition is not a direct transition from the chaotic motion of the mesoscale structural elements to the macroscopic translational motion of the medium as a "noise-induced transition" In the case of a quasi-equilibrium process, the noise-induced transition occurs by means of coupling between the mesoparticle velocity dispersion and the velocity defect (see Chap. 5). In the case of non-equilibrium processes, the transition to a structurally unstable state is driven by the oscillating regime straining the mesoscale. In our investigation, the main focus is on the physical background for the shock-induced structural transition: the stochastic features of the dynamically deformed medium are considered to be the driving forces for such a transition. The transition from one structural scale to another is realised through the excitation of mesoscale oscillations resulting from a back coupling between deformation processes at the macroscale and mesoscale.

Y. Meshcheryakov, *Multiscale Mechanics of Shock Wave Processes*, Shock Wave and High Pressure Phenomena, https://doi.org/10.1007/978-981-16-4530-3_7

7.1 Introduction

One of the important relationships of multiscale deformation has been shown to be the proportionality between the mesoparticle velocity variance and the macroscopic strain rate [1]:

$$D(t) = R\frac{d\varepsilon(t)}{dt}. \tag{7.1}$$

where R is the proportionality coefficient, the meaning of which will be given later. This relationship links mesoscale 1 in the form of the mesoparticle velocity variance, $D(t)$, and the macroscopic strain rate, $\frac{d\varepsilon(t)}{dt}$. Multiplying both sides of Eq. (7.1) by ρC_p yields:

$$\rho C_p D = \rho C_p R\frac{d\varepsilon}{dt} \tag{7.2}$$

The left-hand side of Eq. (7.2) is the spherical component of the stresses initiated by velocity pulsations at the mesoscale 1. The mesoparticle velocity dispersion D^2 or the velocity variance, D, is quantitative characteristics of the velocity pulsations. The value $\eta_{mc} = \rho C_p R$ is the macroscale viscosity. Thus, Eq. (7.2) shows that in a steady wave state, the pressure initiated by the mesoparticle velocity distribution, (i.e. chaotic motions of mesoparticles) is balanced by the viscous stress, which results from the translational motion of the carriers of deformation.

The question arises: What is the physical mechanism of the transition from the chaotic motion of mesoparticles to the translational motion of the dynamically deformed medium?

7.2 The Oscillating Regime of the Dynamically Deformed Heterogeneous Medium

This problem concerns the propagation of a plane shock wave in a relaxing medium. In Chap. 6, a mixed kind of constitutive equation was used for locking the balance equation of a medium. Here, we consider the situation when only the mesoscopic mechanism of stress relaxation is taken into account. We accept that the relaxation of the equilibrium state of the heterogeneous medium flows due to the meso–macromomentum exchange, the quantitative characteristic of which is the velocity defect.

For one-dimensional propagation of shock wave, the balance equations for momentum conservation and medium continuity

$$\rho u_t - \sigma_x = 0; \tag{7.3}$$

$$u_x - \varepsilon_t = 0; \tag{7.4}$$

are constrained by the constitutive equation for a relaxing medium introduced by Duvall [2] and Taylor [3]:

$$\sigma - \rho C_l^2 \varepsilon = -2\mu\varepsilon^p \tag{7.5}$$

or

$$\sigma_t - \rho C_l^2 \varepsilon_t = -F. \tag{7.6}$$

Here C_l is the longitudinal sound velocity, μ is the shear modulus, and F is the relaxation function:

$$F = 2\mu \frac{\mathrm{d}\varepsilon^p}{\mathrm{d}t}. \tag{7.7}$$

Theoretical and experimental investigations show that a direct transition from dislocation dynamics to the macroscopic state is not possible as the additive dislocation mechanism of stress relaxation cannot provide a sufficient strain rate for an adequate description of both the elastic precursor decay [4] and the plastic front [5]. In Chap. 6, as an additional stress-relaxing mechanism, the rate of change of the velocity defect was used in modelling the shock wave processes. In this case, an acceptable agreement between shock wave theory and experiments was obtained. In addition, the theory reveals the possibility of initiating the oscillations at the shock wave front. The oscillations of the shock wave front have previously been found in a series of experiments on shock loading copper [6] and aluminium alloys [7].

In the present chapter, to elucidate the physical nature of the meso–macrotransition, a theoretical investigation of the oscillation regime at the mesoscale is carried out. As distinct from the Duvall–Taylor approach [2, 3] and mixed approach [5], in the present investigation, the stress relaxation in a dynamically deformed medium is realised only through the motion of ECD at the mesoscale. Dislocations and other carriers of deformation at the atom-dislocation scale are considered to provide the formation of mesoparticles during the dynamic deformation. Once formed, the mesoparticles are self-consistent carriers of deformation. In this situation, the stress relaxation in Eq. (7.6) is determined only by the plastic strain rate related to the velocity defect. The advantage of such an approach is that the parameters of the dislocation structure are not incorporated into the relaxation model. The density and velocity of the dislocations cannot be controlled under the conditions of dynamic deformation. As distinct from dislocation structure, the defects of the particle velocity can be measured in real time. We accept that the stress relaxation in Eq. (7.7) flows by changing the velocity defects. By analogy with the one-dimensional shock wave deformation, $\varepsilon = \frac{u}{C_0}$, the contribution to deformation due to velocity defects can be written in the following form:

$$\Delta \varepsilon^p = \frac{\delta u_{def}}{C_0}, \tag{7.8}$$

The velocity defect has been determined in Chap. 5 as follows:

$$\delta u_{\rm def} = \frac{1}{2}\frac{{\rm d}D^2}{{\rm d}u}. \tag{7.9}$$

Equation (7.9) can then be written in the form:

$$\delta u_{\rm def} = \frac{1}{2}\frac{{\rm d}D^2}{{\rm d}u} = D\frac{{\rm d}D}{{\rm d}u} = D\left[\frac{{\rm d}D/{\rm d}t}{{\rm d}u/{\rm d}t}\right]. \tag{7.10}$$

Substituting Eq. (7.10) into Eq. (7.8) yields:

$$\Delta \varepsilon^p = \frac{R^2}{C_0^2}\frac{{\rm d}^2\varepsilon}{{\rm d}t^2}. \tag{7.11}$$

Thus, the relaxation function in Eq. (7.7) can be written as follows:

$$F = \frac{2\mu}{C_l}\varepsilon_{tt}. \tag{7.12}$$

and the normal stress in Eq. (7.5) equals:

$$\sigma = \rho C_l^2 \varepsilon - \frac{2\mu}{C_l^2}\varepsilon_{tt}. \tag{7.13}$$

The equation system (7.3)–(7.4) is reduced to a second-order differential equation:

$$\rho \varepsilon_{tt} - \sigma_{xx} = 0; \tag{7.14}$$

Substituting Eq. (7.13) into (7.14) yields:

$$\varepsilon_{tt} - \rho C_l^2 \varepsilon_{xx} - 2\left(\mu R^2/\rho C_0^2\right)\varepsilon_{xxtt} = 0. \tag{7.15}$$

For the case of the steady propagation of a shock wave, the single variable $\zeta = x - C_p t$ can be used, and Eq. (7.15) takes the form:

$$(C_l^2 - C_p^2)\varepsilon_{\varsigma\varsigma} + 2\left(\mu R^2/\rho C_l^2\right)\varepsilon_{\varsigma\varsigma\varsigma\varsigma} = 0. \tag{7.16}$$

Exchanging $\frac{d^2\varepsilon}{d\varsigma^2} = \psi$ leads to the equation for the oscillator:

$$(C_l^2 - C_p^2)\psi + 2\left(\mu R^2/\rho C_l^2\right)\Psi_{\varsigma\varsigma} = 0. \tag{7.17}$$

The frequency of the oscillations equals

$$\omega = \frac{1}{2}\frac{\rho C_l^2(C_l^2 - C_p^2)}{\mu R^2}. \tag{7.18}$$

Physically, the oscillatory regime of shock wave propagation is a sequence of positive back coupling provided by Eqs. (7.11) and (7.13), which are used for constraining the balance Eqs. (7.3) and (7.4). The positive back coupling between the macroscale and the mesoscale occurs between the particle velocity dispersion and velocity defect.

In our experiments, the size of the laser spot of the interferometer (50–70 μm) at the free surface of the target corresponds to mesoscale 2 (50–500 μm); this experimental technique allows the recording of (i) the motion of a single structural element of mesoscale 2 as a whole and (ii) the high-frequency oscillations at mesoscale 2. Where observed, the oscillations are excited at the top of the plastic front.

Figure 7.1 demonstrates velocity oscillations with a period of 2.5 ns in a 7 mm 1565 aluminium alloy target loaded at the impact velocity of 636.5 m/s. Analogous oscillations with a period of 1.2 ns were registered in M3 copper (Fig. 7.2).

In 30CrNi4Mo steel, the oscillations are suppressed by the strong dissipative effect. Instead of oscillations, there is a short-duration decrease in particle velocity after which the signal is restored (Fig. 7.3).

The oscillations are initiated at the impact velocity, which is higher than the threshold for the structurally unstable transition, U_{ins}. Although the oscillations are initiated due to the interaction of stochastic features of the dynamically deformed structure in the form of the particle velocity dispersion, the meso–macrotransition itself happens due to the swinging high-frequency oscillations of the mesostructure. Thus, it may be concluded that the meso- to macrostructural transition is not a direct transition from the chaotic motion of the mesoscale structural elements to the macroscopic translational motion of the medium as a "noise-induced transition" [8]. In the case of a quasi-equilibrium process, the noise-induced transition occurs by means of coupling between the mesoparticle velocity dispersion and the velocity defect (see Chap. 5). In the case of non-equilibrium processes, the transition to a structurally unstable state is driven by the oscillating regime straining the mesoscale.

Recording the oscillation structure of the plastic front allows proportionality coefficient R in Eq. (7.1) to be calculated. We use, firstly, Eq. (7.18) for oscillator frequency and, secondly, the experimental profile of the free surface velocity in the 1565 aluminium alloy where the oscillations have been registered. (Fig. 7.1). It follows from Eq. (7.18):

$$R^2 = \frac{\rho C_l^2}{\mu\omega^2}\left(C_l^2 - C_p^2\right). \tag{7.19}$$

The physical characteristics and parameters of the shock wave for 1565 aluminium alloy are provided in Table 7.1.

Here C_l is the velocity of the elastic wave, C_p is the velocity of the plastic front, T_{fr} is the rise time of the plastic front, T_{osc} is the period of oscillations at the top

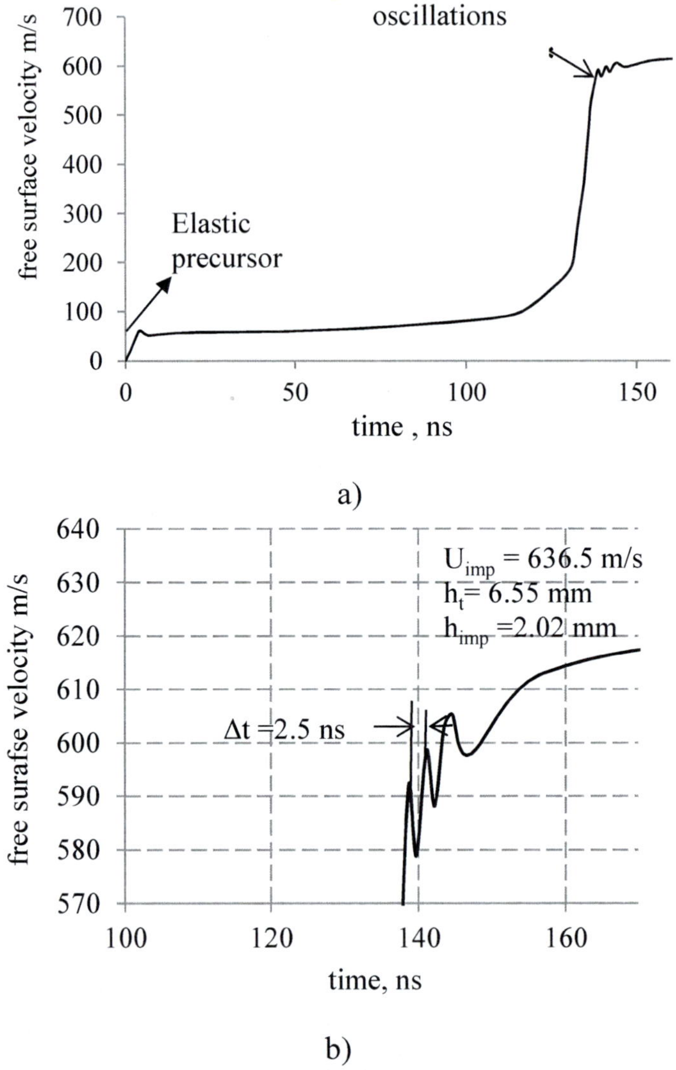

Fig. 7.1 **a** Oscillations at the top of plastic front in 1565 aluminium alloy target at the impact velocity of 636.5 m/s, **b** increased pattern of oscillations

of the plastic front, and μ is the shear modulus. Calculation using Eq. (7.19) results in R = 48.7 μm, which coincides with the spatial duration of the plastic front L_f = 49.77 μm.

In Fig. 7.4, the microstructural pattern for the shock-induced plastic flow oscillations in the form of periodically plastically deformed regions in shock-loaded aluminium A995 is presented.

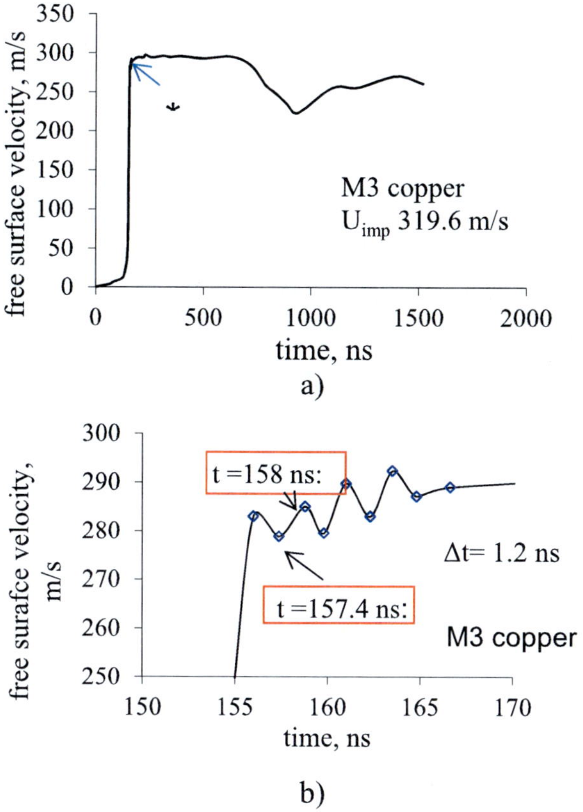

Fig. 7.2 **a** Oscillations at the top of plastic front in M3 copper at the impact velocity of 319 m/s, **b** increased pattern of oscillations

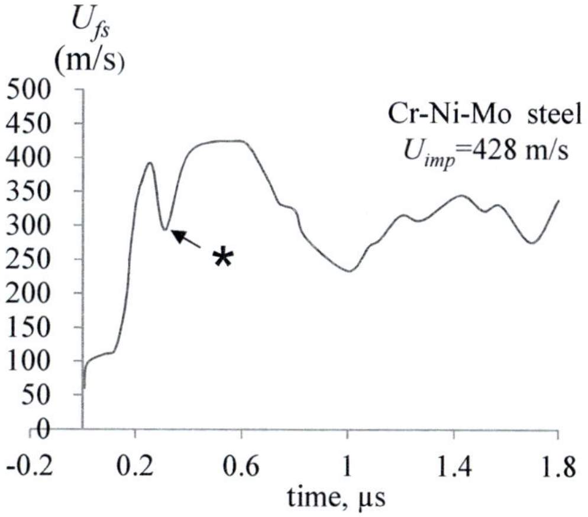

Fig. 7.3 Short-duration loss of structural stability in 30CrNi4Mo steel target. The material becomes structurally unstable for 0.18 μs when the free surface velocity decreases by ~107 m/s

Table 7.1 Parameters of elastic-plastic wave and shear modulus for 1565 aluminum alloy

Parameters	C_l mm/μs	C_p mm/μs	T_{fr} ns	T_{osc} ns	μ GPa	L_{osc}μm
Al. 1565	6.37	5.53	9.0	9.0	27	49.77

Fig. 7.4 Periodic plastic flow initiated in an A995 aluminium target shock-loaded under uniaxial strain conditions

Consider now the spectral features of Eq. (7.20), supposing that the dynamic deformation can be written in the harmonic form:

$$\varepsilon(x, t) = Q \exp(kx - \omega t). \tag{7.20}$$

Equation (7.18) can be transformed into the dispersion equation:

$$\omega^2 - C_l^2 k^2 = \frac{\mu R^2}{\rho C_l^2} k^2 \omega^2. \tag{7.21}$$

where

$$\omega = \frac{C_l k}{\sqrt{1 + \frac{\mu k^2 R^2}{\rho C_l^2}}}. \tag{7.22}$$

The frequency of oscillations decreases when the spatial duration of the plastic front R increases. The plastic wave velocity equals:

$$C_{pl} = \frac{\omega}{k} = \frac{C_l}{\sqrt{1 + \frac{\mu k^2 R^2}{\rho C_l^2}}}. \tag{7.23}$$

When $R \to 0, C_{pl} \to C_{l.}$; that is, the plastic wave velocity equals the longitudinal sound velocity. In this limit, the elastic–plastic wave is known to be transformed into a discontinuous wave. In contrast, when $R \to \infty$, $C_{pl} \to 0$; that is, wave motion is absent.

The results of the present chapter do not contradict the synergetic approach developed in [8], where the transition to the new form of medium state is considered a "noise-induced transition". In our investigation, the main focus is on the physical background for the shock-induced structural transition: the stochastic features of the dynamically deformed medium alone are considered to be the driving forces for such a transition. The transition from one structural scale to another is realised through the excitation of mesoscale oscillations resulting from a back coupling between deformation processes at the macroscale and mesoscale.

References

1. Indeitzev DA, Meshcheryakov YuI, Kuchmin AYu, Vavilov DA (2015) Multiscale model of steady wave shock in medium with relaxation. Acta Mech 22(3):917–930
2. Duvall GE (1964) Propagation of plane shock waves in a stress-relaxing medium. In: Kolsky H, Prager W (eds) Stress Waves in Inelastic Solids. Berlin, Springer, pp 20–32
3. Taylor JW (1965) Dislocation dynamics and dynamic yielding. J Appl Phys 36(10):3146–3155
4. Johnson JN, Jones OE, Michaels TE (1970) Dislocation dynamics and single-crystal constitutive relation: shock-wave propagation and precursor decay. J Appl Phys 41(6):2230–2239
5. Meshcheryakov YI (2016) Particle velocity non-uniformity and steady-wave propagation. Shock Waves Detonations Explos 26(3) https://doi.org/10.1007/S0093-016-0659-7
6. Meshcheryakov YuI, Divakov AK, Zhigacheva NI, Makarevich IP, Barakhtin BK (2008) Dynamic structures in shock-loaded copper. Phys Rev B 78:64301–64316
7. Meshcheryakov YuI, Divakov AK, Zhigacheva NI, Konovalov VG, Osokin EP (2019) Formation of multiscale structure under shock loading. Phys Mesomech 22(3):44–53
8. Horsthemke W, Lefever R (1984) Noise-Induced transitions. Spinger. Berlin. Heidelberg, New York, Tokyo, p 380

Part II
Mesoscale Approach to the Dynamic Properties of Materials

Chapter 8
Experimental Techniques for Shock Loading

Abstract Three experimental techniques are used in the present investigation: (i) shock loading under uniaxial strain conditions (planar collision of an impactor and a target), (ii) pulse loading of planar targets with a high-power electron beam and (iii) the penetration of elongated hard rods into a planar target.

(i) Tests conducted under uniaxial strain conditions
(ii) For short times, near the loaded contact plane, the effects of viscosity and thermo-conductivity are usually very small. During loading with short stress pulses, the slow diffusion processes that provide the transport of momentum and energy do not have time to develop. In this situation, investigation of the deformation processes in the contact region of the shock-loaded target is possible only by loading with stress pulses of very short duration. In the present investigation, loading with short stress pulses was performed using a high-power pulse electron beam.
(iii) The penetration tests allow the structure to evolve depending on the impact velocity to be retraced by means of parallel structural investigations of specimens that experience different shock loading schemes.

8.1 Introduction

As developed in previous chapters, the concept of the dynamic mesoscale proves to be necessary to define the criteria characteristics of the shock wave behaviour of solids. In the case of high-velocity deformation, the determination of the criteria characteristics at the mesoscale demands the development of specific experimental techniques. The complexity of investigating shock wave processes relates to the requirement that interpretations obtained from experimental shock data must account for the natural laws of shock wave propagation. However, our theoretical models include the experimental data sought. The development of adequate theoretical models is also difficult because of the absence of precise spatio-temporal profiles for the shock waves. The published experimental data cannot often be used for comparison with theoretical models because the experimental conditions are not clear. The most reliable data are

Y. Meshcheryakov, *Multiscale Mechanics of Shock Wave Processes*, Shock Wave and High Pressure Phenomena, https://doi.org/10.1007/978-981-16-4530-3_8

thought to be obtained in investigations where the following three components are present:

- Shock wave experiments that record wave profiles,
- Microstructural investigations of post-shock specimens,
- Theoretical modelling of high-velocity processes.

In Part 1, the conditions for shock-induced structural instability have been theoretically investigated. Instead of previous models for instability initiation such as the thermal model of Asay and Grady [1–3], the interference of tension wave approach of Buravova [4], the localised wave interference model of Khantuleva-Meshcheryakov [5], in the present study, we suggest the "local kinetic" approach, based on the mesoscale velocity distribution. This approach is grounded in abundant experimental data on the localised measurement of the response of solids to an impact by using a high-resolution interference technique. This technique allows both the mean particle velocity, the first statistical moment of the particle velocity distribution function (PVDF), and the mesoparticle velocity dispersion, the second statistical moment of the PVDF, to be measured in real time.

Distinct from the quasistatic strain rates, both the dynamic yield limit and dynamic stress–strain diagram change during the loading in the microsecond regime of shock loading. Here, the kinetics of the internal structure plays a dominant role. Specifically, the propagation and decay of stress pulses are sensitive to and dependent on the velocity distribution of the ECD at several scales, which require the formulation of a multiscale kinetic theory. As a rule, at one scale level, the carriers of dynamic deformation are in an equilibrium state, whereas at another scale level, the kinetics of the carriers is far from the equilibrium state.

In the present study, the diversity of techniques for shock loading is not considered. The methods for shock loading are sufficiently elucidated elsewhere in the literature. Three experimental techniques are used in the present investigation: (i) shock loading under uniaxial strain conditions (planar collision of an impactor and a target), (ii) pulse loading of planar targets with a high-power electron beam and (iii) the penetration of elongated hard rods into a planar target.

8.2 Shock Loading Under Uniaxial Strain Conditions

In order to study the link between the macroscopic response to shock loading and the microstructural behaviour of a material, the geometry of loading must be as simple as possible, so the macroscopic 3D-effects of the wave interactions are excluded, at least during the time of measurement. In this scenario, tests conducted under uniaxial strain conditions are thought to be the most informative since in this configuration, the shock front shape is highly sensitive to the structural characteristics of the material. Moreover, these tests provide information on the resistivity of a material to both compression and tensile stresses. Tests under uniaxial strain conditions also allow the strain rate dependence of the material to be studied in the simplest possible

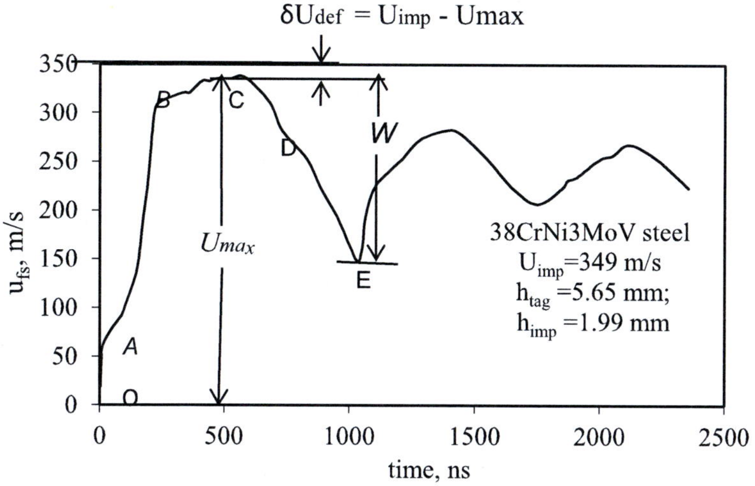

Fig. 8.1 Free surface velocity profile for a 5 mm 38CrNi3MoV steel target loaded at an impact velocity of 349 m/s

manner. For example, from the point of view of the elastic precursor behaviour, the strain rate dependence is revealed first, by the decrease of stress at the elastic precursor near the loaded surface; and second, by the valuable (2–3 times) increase in the steady state value of the amplitude of the elastic precursor. As for the plastic front, the strain rate dependence is revealed by the fact that the propagation velocity is dependent on the loading conditions.

Unlike the experiments with rods, the tests under uniaxial strain conditions allow, in addition to the dynamic yield limit, the spall strength and the threshold of the structural instability under dynamic compression to be recorded in real time. (see Chap. 10). The typical free surface profile registered during shock loading of a 5 mm 38CrNi3MoV steel target at an impact velocity of 349 m/s is shown in Fig. 8.1.

Let us interpret the separate pieces of the velocity profile in Fig. 8.1. Piece *OA* corresponds to the elastic precursor, piece *AB* is the plastic front of the shock wave, and $U_{\max}$ is the maximum value of the free surface velocity at the plateau of the compressive pulse. The difference in velocities, W, between the maximum velocity (point C) and the first minimum at the back front (point *E*) is called the "pull-back velocity" and corresponds to the spall strength of the material.

A schematic of the shock test facility, together with the equipment for measuring the response of the material on impact, is presented in Fig. 8.2. It includes, along with the gas gun, a two-channel laser interferometer and the arrangement for independent measurement of the impactor velocity. The target (3), in the form of a disc, is mounted at the output of the barrel (6), inside the vacuum chamber (1); the striker consists of a metallic impactor and a sabot manufactured from polyvinyl carbonate (see Fig. 8.3).

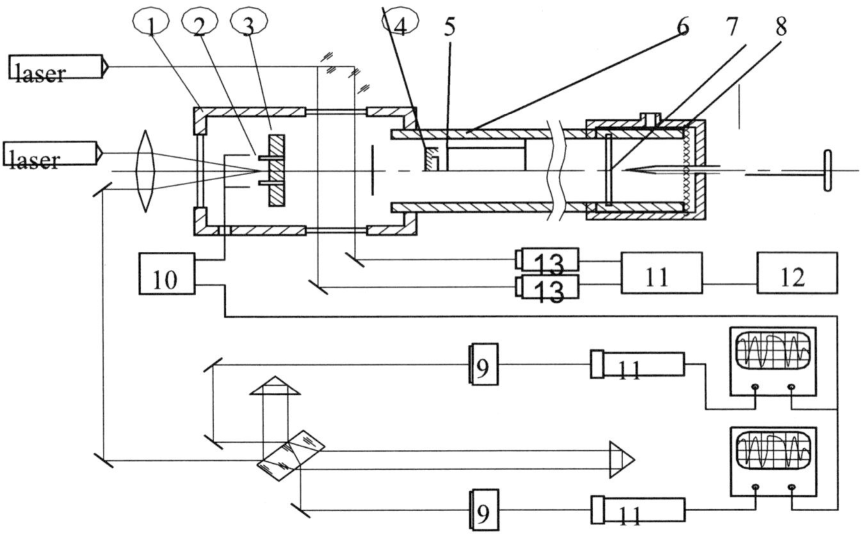

1 – vacuum chamber, 2 - pins, 3 – target, 4 - impactor, 5.- sabot, 6 – barrel, 7 - membrane, 8 – high pressure chamber, 9 - polaroid, 10 - detectors, 11-12 – amplifiers, 13 – equipment for measuring the time intervals.

Fig. 8.2 Schematic of the facility for shock tests under uniaxial strain conditions

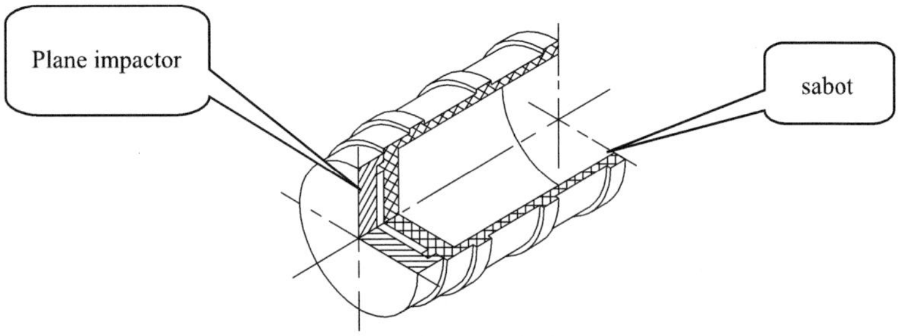

Fig. 8.3 Impactor and sabot for tests under uniaxial strain conditions

8.3 The Pulse Loading of Plane Targets with a High-Power Electron Beam

One unusual and interesting phenomenon revealed in shock wave experiments is the anomalous loss of shock wave amplitude, which cannot be explained by the direct transformation of mechanical energy into heat through dissipation. This phenomenon is present in two situations: (i) during the transition of the plastic front into the plateau of the compressive pulse and (ii) within the contact region of the shock-loaded target. The first kind of particle velocity loss takes place at impact velocities higher than the structural instability threshold. The threshold impact velocity corresponds to the situation where the mesoscale 1 relaxation processes are exhausted and new, large-scale structures are nucleated during the process of transition of material to the next structural scale (mesoscale 2).

The second situation has a different physical nature. For short times, near the loaded contact plane, the effects of viscosity and thermo-conductivity are usually very small. During loading with short stress pulses, the slow diffusion processes that provide the transport of momentum and energy do not have time to develop. Specifically, in shock wave experiments and theoretical modelling, the effects of thermo-conductivity and diffusion are neglected. In this situation, investigation of the deformation processes in the contact region of the shock-loaded target is possible only by loading with stress pulses of very short duration. In the present investigation, loading with short stress pulses was performed using a high-power pulse electron beam. Two tasks arise from pulsed electron beam loading: (i) the generation of short stress pulses of nanosecond duration; (ii) the study of the processes of dynamic deformation under the condition of the stress pulses of high amplitude and short duration. As direct measurement of the pressure or the particle velocity at the radiated surface is impossible, the only indication is the determination of these values at the free surface of targets of different thicknesses and extrapolation to the loaded surface of the target. The extrapolation of stress pulses into the region of shallow depths yields the problem of calculating stress pulses at shallow depth following the known decay law of stress pulses during propagation through the target. In other words, to extrapolate the pressures at shallow depth, the theoretical decay dependence must have previously been determined. However, the determination of the theoretical dependence for the decay of the elastic–plastic wave is a separate problem of dynamic plasticity. One specific feature of the propagation of short stress pulses is hydrodynamic decay. This kind of decay results from the fact that the velocity of the back front is higher than the velocity of the forehead front of the compressive pulse. Therefore, the back front overtakes the forehead front and decreases the stress pulse amplitude. In the present investigation, to determine the stress values at a small distance from the loaded surface, we use the fact that, unlike the plastic wave, the elastic precursor does not suffer from hydrodynamic decay. Its initial amplitude coincides with the amplitude of the plastic wave before the separation of the shock wave by the elastic precursor and the plastic wave. For longer time intervals, the elastic precursor propagates at sound velocity away from the plastic front. Thus, for the determination of the initial amplitude of the stress pulse at the loaded surface of the target, one can use the extrapolation of dependence for the elastic precursor decay.

In our experiments, to determine the elastic precursor decay law, we use the experimental curve for D16 aluminium alloy target measured with the high-power electron beam accelerator "Neva-01" of the Saint-Petersburg Institute of Electro-Physical Apparatuses. The physical characteristics of this accelerator are provided in Table 8.1.

Table 8.1 Physical characteristics of the high-power electron beam accelerator "Neva-01"

Electron energy	800 keV
Electron beam current	200 kA
Electron pulse duration	70 ns
Total energy of electron beam	10 kJ

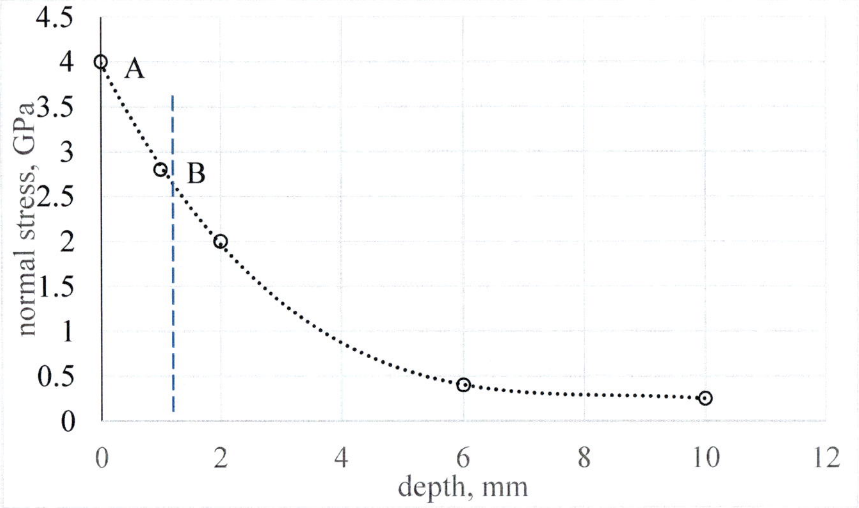

Fig. 8.4 Decay of the elastic precursor in D16 aluminium alloy. The dependence has been obtained by loading with the pulse electron beam

The curve of the elastic precursor decay measured in the process of pulse loading targets of different thicknesses is provided in Fig. 8.3. The extrapolation of this dependence to the collision plane allows the initial stress to be determined in the case of a loading of nanosecond duration. For example, in D16 aluminium alloy, the initial stress at the contact surface of the target equals 4 GPa (see Fig. 8.4). The dotted line indicates the minimum target thickness for our experiments. The line segment AB is the result of extrapolation.

8.4 The Penetration of Elongated Hard Rods into Plane Target

Experiments on the penetration of elongated rods were conducted at the same facility as the uniaxial strain condition tests. The only characteristic measured in these tests is the impact velocity. In parallel, for each impact velocity, the planar tests were performed, which allows the correlation of the transient processes in the two loading schemes.

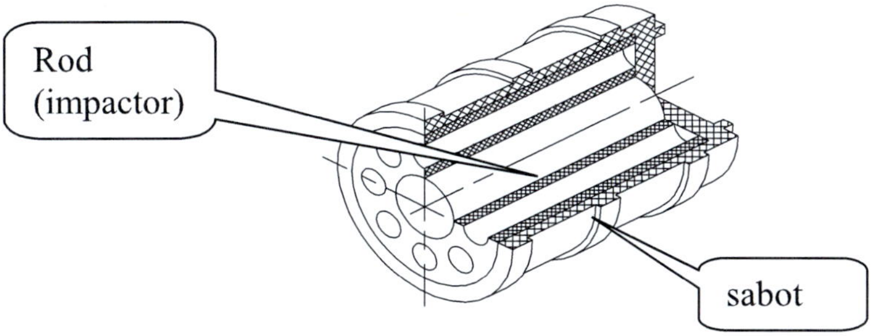

Fig. 8.5 Polyvinyl carbonate sabot and impactor rod

To ensure the rod is perpendicular to plane of the target, the 20 mm long rod of 5 mm diameter is mounted into a polyvinyl carbonate sabot (Fig. 8.5). The conditions for a "rigid rod and target" are provided by using the high-strength 02Cr18Co9Mo5-VI maraging steel to make the rod.

The penetration tests allow the structure to evolve depending on the impact velocity to be retraced by means of parallel structural investigations of specimens that experience different shock loading schemes. The cross sections of post-shock targets were studied with an Axio-Observier Z-1 m microscope after polishing and etching.

References

1. Grady DE, Asay JR (1982) Calculation of thermal trapping in shock deformation of aluminum. J Appl Phys 53(1):7350–7354
2. Grady DE, Kipp ME (1987) The growth of unstable thermoplastic shear with application to steady-wave shock compression in solid. 35(1):95–119
3. Glenn LA (1988) Strain-energy effects on dynamic fragmentation. J Appl Phys 59(4):15
4. Belikova AF, Buravova CN, Petrov EV (2013) Localization of deformation under dynamic loads. J Tech Phys 83(8):68. (Russian journal)
5. Khantuleva TA, Meshcheryakov YI (2015) Non-equilibrium processes in condensed media. Part 2. Shock-induced structural instability. Phys Msomech 18(1):14

Chapter 9
How to Measure the Parameters of Mesoscale

Abstract Data on the dynamic strength and plasticity of the material, including the dynamic yield limit, spall strength, threshold of structural instability and particle velocity distribution, are inferred from the temporal profiles of the free surface velocity, $u_{\mathrm{fs}}(t)$, and velocity dispersion profiles which are registered with the velocity interferometer. Three dynamic characteristics are used in the present analysis. One of the basic characteristics of dynamic deformation that reflects the transition of a material into a structurally unstable state and provides quantitative data on the exchange of energy and momentum between scales is the "defect of particle velocity". It is determined in tests under uniaxial strain conditions as the difference between the velocity of the impactor under a symmetrical collision and the maximum free surface velocity at the plateau of the compressive pulse, $\delta u_{\mathrm{def}} = (U_{\mathrm{imp}} - U_{\mathrm{fs\ max}})$.

The second dynamic characteristic is the threshold of structural instability. It characterises the beginning of shock-induced structural heterogenization of a material. This characteristic can be obtained from the series of free surface velocity profiles measured under uniaxial strain conditions over a range of impact velocities. The maximum free surface velocity, $U_{fs\ \max} = f(U_{\mathrm{imp}})$, at which the velocity defect begins to increase drastically is defined as the "threshold of structural instability", U_{ins}. Last, the third dynamic strength characteristic determined in tests under uniaxial strain conditions is the "spall strength or pull-back velocity".

9.1 Introduction

Developed in Chap. 2, the theoretical approach demonstrates the stochastic nature of elementary carriers of deformation (ECD) at the mesoscale. The totality of ECD at the mesoscale is shown to be short-lived velocity pulsations [1, 2]. The union of the mean particle velocity and the velocity pulsations results in the "wavy shape" of the summary shock front. Both experimental and theoretical studies indicate heterogeneous non-equilibrium particle velocity in the shock compression of solids. Specifically, the theoretical description of the wavy front in polycrystalline materials applies the concept of random changes in the direction of shock waves in differently oriented

Y. Meshcheryakov, *Multiscale Mechanics of Shock Wave Processes*, Shock Wave and High Pressure Phenomena, https://doi.org/10.1007/978-981-16-4530-3_9

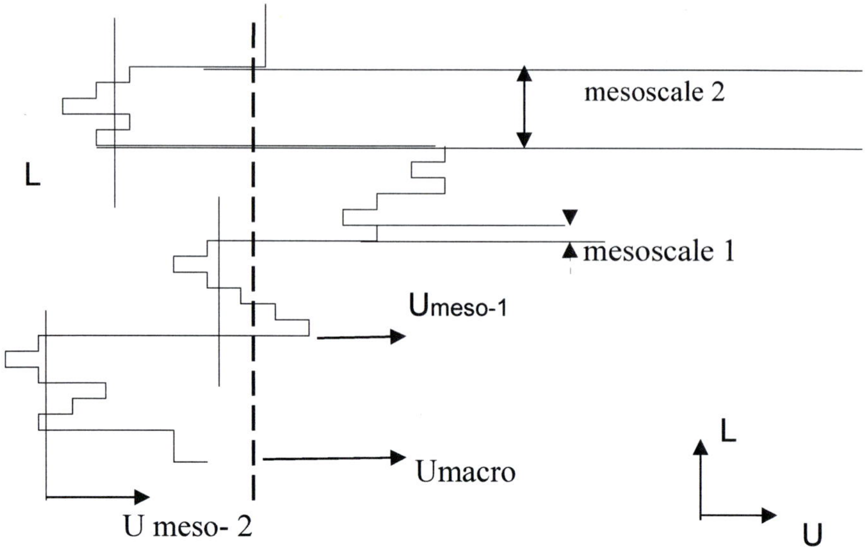

Fig. 9.1 *U-L* velocity-space multiscale configurations of a shock front in a heterogeneous medium

grains of polycrystalline solids. Meyers and Carvalho investigated shock wave propagation in a polycrystalline solid (nickel) using a probabilistic model for crystal orientations [3]. It was found that the shock front becomes irregular as it propagates into the material. An analogous investigation has been performed by Case and Horie for polycrystalline copper [4]. These investigations contradict the conventional continuum model of shock wave structure [5].

In Fig. 9.1, a qualitative picture of the random configuration of such a wavy front in a heterogeneous medium is presented in the form of a meandering line, and three length scales are presented: mesoscale 1, mesoscale 2 and macroscale. The dotted line *NN'* corresponds to the mean velocity of the shock front (macroscale), while the meandering lines represent the random velocities of separate pieces of the shock front at mesoscale 1 and mesoscale 2.

The technique for recording the mean particle velocity (the first statistical moment of the PVDF) has been previously developed [6–9]. This technique has been modified in several directions, the last of which, the VISAR interferometer, can be used to determine the free surface velocity temporal profiles for any reflecting surfaces [10]. It is important to emphasise that VISAR can register the time history of only the mean particle velocity.

The next step in registering the response of solid to an impact is the development of an interference technique which is capable of measuring the other parameters, the particle velocity distribution function (the second statistical moment of the PVDF) and the mesoparticle velocity dispersion [11, 12]. In this technique, it is expected that the particle velocity distribution function is at equilibrium, so it can be characterised with only two statistical moments, the mean particle velocity, u(t), and the particle velocity dispersion, $D^2(t)$.

9.2 Experimental Techniques

Shock tests under uniaxial strain conditions were conducted using a light gas gun. The planar targets were discs 52 mm in diameter and 7 mm thick. Data on the dynamic strength and plasticity of the material, including the dynamic yield limit, spall strength, threshold of structural instability and particle velocity distribution, are inferred from the temporal profiles of the free surface velocity, $u_{fs}(t)$, and velocity dispersion profiles which are registered with the velocity interferometer developed in our study. To date, this kind of interferometry is the only experimental technique which allows the recording of both the mean free surface velocity and the particle velocity distribution in real time. The quantitative characteristic of the particle velocity distribution is the velocity variance. It should be noted that the laser spot at the free surface of the target is approximately equal to or smaller than a mesoscale 2 structural element (50–100 μm). Thus, the fringe signal of the interferometer corresponds to the response of a single mesoscale 2 structural element, while the velocity variance corresponds to the velocity distribution inside a single mesoscale 2 element. To obtain the macroscopic free surface response of the target on impact, one should average over the mesoscale 2 responses. The free surface velocity profile shown in Fig. 9.2 is a typical response of a single structural element for metals deformed under

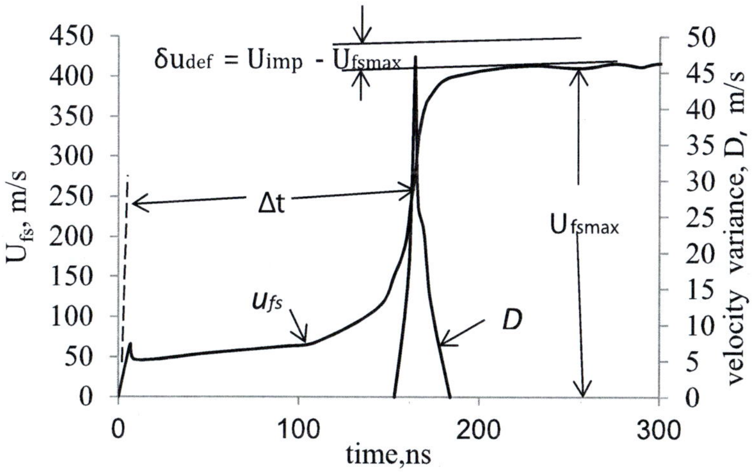

Fig. 9.2 Free surface velocity profile, $u_{fs}(t)$, and the particle velocity variance profile, $D(t)$, for a 1561 aluminium alloy target at an impact velocity of 442.5 m/s

uniaxial strain conditions. This profile provides the mean free surface velocity, $u_{fs}(t)$, at mesoscale 2 and the velocity variance, $D(t)$, at mesoscale 1 (1–10 μm).

Three dynamic characteristics are used in the present analysis. One of the basic characteristics of dynamic deformation that reflects the transition of a material into a structurally unstable state and provides quantitative data on the exchange of energy and momentum between scales is the "defect of particle velocity". It is determined in tests under uniaxial strain conditions as the difference between the velocity of the impactor under a symmetrical collision and the maximum free surface velocity at the plateau of the compressive pulse, $\delta u_{def} = (U_{imp} - U_{fs\ max})$ (Fig. 9.2).

The second dynamic characteristic is the threshold of structural instability. It characterises the beginning of shock-induced structural heterogenization of a material. This characteristic can be obtained from the series of free surface velocity profiles measured under uniaxial strain conditions over a range of impact velocities. The maximum free surface velocity, $U_{fs\ max} = f\left(U_{imp}\right)$, at which the velocity defect begins to increase drastically is defined as the "threshold of structural instability", U_{ins}.

Last, the third dynamic strength characteristic determined in tests under uniaxial strain conditions is the "spall strength or pull-back velocity" which is determined in accordance with [13].

9.3 Analysis of the Velocity Interferometer Under Conditions of Mesoparticle Velocity Distribution.

Presented in Fig. 9.2, the free surface velocity profile has been recorded with the velocity interferometer illustrated in Fig. 9.3. A laser beam reflected from the free surface of target 1 is subdivided into two beams by semi-transparent mirror 3. After travelling the different routes *BGD* and *BJID*, the beams experience interference on mirror 4. The difference in the lengths of *BGD* and *BJID* presents the delay shoulder of the velocity interferometer, l_d. The delay shoulder is the element of the interferometer where the fringe signal is generated. Detailed analysis of a velocity interferometer results in the following formula which relates the number of beat frequency of the fringe signal and the velocity of the free surface of the target [7]:

$$u(t) = \frac{\lambda_0}{2\tau_d} N(t). \tag{9.1}$$

By counting the number of beat frequencies of the fringe signal over the finite time interval $N(t)$, it is possible to plot the time dependence of the free surface velocity $u_{fs}(t)$, that is, the time-resolved velocity profile. The relative error in the determination of the instant fringe beat frequency $\Delta F(t)$ equals τ_d/t_{ch}, where t_{ch} is the time in which the velocity increases from zero to its maximum value. In [8, 9], the basic requirements of the adjustable parameter of the velocity interferometer, τ_d, are determined:

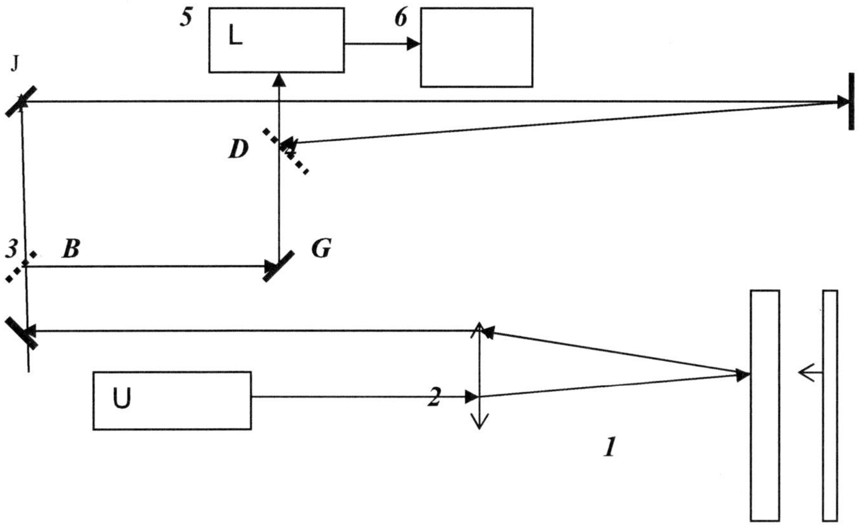

Fig. 9.3 Schematic of a velocity interferometer: 1—laser, 2—lens, 3 and 4—semi-transparent mirrors, 5—register, 6—oscilloscope

1. The delay time should be chosen from the conditions such that the fringe signal can be reliably detected with electronic equipment

$$\frac{2\tau_d}{\lambda_0}\dot{u}(t) \le 10^7\,\mathrm{C}^{-1}, \tag{9.2}$$

2. For measuring of the free surface velocity profile with an error velocities for a concrete moment.

In reality, the velocities of separate particles over the transverse cross section of the laser spot at the free surface can be different due to the heterogeneity of the dynamic deformation process. If the size of the elementary velocity non-uniformity at the free surface of the target is much greater than the size of the laser spot, the measured velocity corresponds to the macroscopic response of the target surface. However, if the size of the velocity non-uniformity is smaller than the diameter of the laser spot but greater than that of the laser radiation, the laser radiation reflected from the free surface can be considered to consist of elementary beams, each of which undergoes an individual Doppler shift during the interaction with the target. The value of each shift is determined by the velocity of the mesoparticle from which the micro-beam is reflected. When reflected from the free surface of the target, the micro-beams pass through the interferometer, where each micro-beam generates its own beat frequency at the output of the interferometer. The resulting signal is averaged over the different beat frequency signals, which results in a decrease in interference contrast. Thus, the degree of contrast of the interference signal directly relates to the value of the mesoparticle velocity distribution. This described effect underpins the quantitative method of measuring the particle velocity dispersion. In the case of the particle velocity distribution at the free surface of the target, the amplitude of the laser radiation reflected from the free surface relates to the number of beat frequencies as follows:

$$I = \frac{1}{2} I_0 \int_{-\infty}^{\infty} [1 + \sin(ku + \delta)] f(u - u_0) \mathrm{d}u.$$

where $f(u - u_0)$ is the PVDF.

Consider now the situation when the beat frequency has lost mono-chromaticity and is characterised by the radiation spectrum Δw. It is well-known that if the light is not monochromatic and if it has a continuous spectrum, the intensity of the radiation over the frequency interval from w to $w + \mathrm{d}w$ is proportional to dw:

$$\mathrm{d}I_o = F(w)\mathrm{d}w \tag{9.3}$$

where $F(w)$ is the spectral density of radiation. Normalisation of $F(w)$ yields:

$$\mathrm{d}I_o = I_o f(w)\mathrm{d}w \tag{9.4}$$

and

$$I_o = \int F(w)\mathrm{d}w \tag{9.5}$$

is the summary radiation density over the transverse cross section of the laser beam and

$$f(w) = F(w) \bigg/ \int F(w)\mathrm{d}w. \tag{9.6}$$

Owing to the interference process, the intensity of the radiation at the output of the interferometer changes in accordance with the expression:

$$\mathrm{d}I = \int \frac{I_0}{2} f(w)(1 + \cos\, w\, \tau_\mathrm{d})\mathrm{d}w \tag{9.7}$$

where $\tau_\mathrm{d} = l_o/c$ is the delay time of the velocity interferometer. The full intensity equals:

$$I = \frac{1}{2} f(w)(1 + \cos\, w\tau\, t_\mathrm{d}) \tag{9.8}$$

As a rule, the radiation spectrum is concentrated in the vicinity of the central frequency $\boldsymbol{w}$, corresponding to the maximum radiation density. Let us exchange integration variables in Eq. (3.9):

$$w = \boldsymbol{w} + w_* \tag{9.9}$$

and take into account that $\Delta w \ll w$. Then one obtains:

$$\mathrm{d}I = (I_o/2)\Phi(w_*)(1 + \cos[(w_* + w)\tau_d]\,\mathrm{d}w_* \tag{9.10}$$

where

$$\Phi(w_*) = f(w_* + \boldsymbol{w}) \tag{9.11}$$

Equation (9.10) can be written in the form:

$$I = (I_o/2)[1 + C\,\cos(\boldsymbol{w}\tau_d) - S\,\sin(\boldsymbol{w}\tau_d)]. \tag{9.12}$$

where

$$C = \int \Phi(w_*)\cos(w_*\tau_d)\mathrm{d}w_* \tag{9.13}$$

$$S = \int \Phi(w_*)\sin(w_*\tau_d)\mathrm{d}w_* \tag{9.14}$$

According to the definition, the contrast of interference signal equals:

$$K = (I_{\max} - I_{\min})/(I_{\max} + I_{\min}) \tag{9.15}$$

From Eqs. (9.13)–(9.15), one obtains:

$$K = \sqrt{(C^2 + S^2)} \tag{9.16}$$

For further analysis, one should determine the concrete shape of the radiation spectrum $\Phi(w_*)$. For a Gaussian-shaped radiation spectrum,

$$\Phi(w_*) = \frac{1}{\sigma\sqrt{2\pi}}\exp(-w_*^2/2\sigma^2)$$
$$\Phi(w_*) = \Phi(-w_*) \tag{9.17}$$

$$K = C = \frac{1}{\sigma\sqrt{2\pi}} \int_{-\infty}^{\infty} \exp(-w_*^2/2\sigma^2)\cos(w_*\tau_d)dw_*$$

or

$$\mathrm{K} = \exp\,(-\tau_d^2\sigma^2)/2 \tag{9.18}$$

where σ is the half-width of the spectral function $\Phi(w)$. Taking into account that

$$\boldsymbol{w} = w_o(1 + 2v/c) \text{ and } w = w_o(1 + 2u/c),$$

one obtains

$$\begin{aligned} w_* &= (4\pi/\lambda_o)(u - v) \\ \sigma &= (2\pi/\lambda_o)D \end{aligned} \tag{9.19}$$

Here, u is the mean particle velocity at the free surface of the target, $\lambda_0 = 2\pi c/w_o$ is the wavelength of the laser radiation, D is the half-width of the particle velocity distribution function or mean square deviation (velocity variance).

The desired relationship between the contrast of the interference signal K and the velocity variance D can be obtained using Eqs. (9.18) and (9.19):

$$K = \exp\left[-\frac{1}{2}\frac{\pi^2 D^2}{U_{\mathrm{int}}^2}\right], \tag{9.20}$$

where $U_{\mathrm{int}} = \lambda_o/2\tau_d$ is the interferometer constant which depends on the wavelength of the laser radiation and the delay time of the velocity interferometer (the shoulder *IDJ* in Fig. 9.3). Then, the intensity of the laser radiation at the output of the interferometer can be written as:

$$I(t) = \frac{I_0}{2}\left\{1 + K\,\cos\left[2\pi\frac{u(t)}{U_{\mathrm{int}}} + \varphi_o\right]\right\}, \tag{9.21}$$

where $\varphi_o = w_o\,\tau_d$ is the initial phase of the fringe signal. The value of the beat frequency of the fringe signal is determined by the mean velocity of the free surface of the target, $u(t)$, whereas the amplitude of the beat frequency depends on the velocity variance $D(t)$. It is important that both characteristics can be determined during the single act of shock loading from the time-resolved free surface velocity profile.

9.4 Investigation of Shock Wave Processes Using the Interference Technique

The above results have been checked by a series of shock-loaded targets of different materials, such as aluminium and its alloys, steels and copper. These materials have a wide spectrum of properties, including different degrees of initial and shock-induced non-homogeneity. One of the example of interference signals measured by the velocity interferometer for a 5 mm steel target is provided in Fig. 9.4. In this figure, the number of beats of the fringe signal at the front of the compressive pulse equals 1.5. For this test, the interference constant, $U_{\text{int}} = 100$ m/s, by one beat, so the value of the free surface velocity equals 150 m/s. This value coincides with the impact velocity and satisfies the well-known free surface approximation rule:

$$u_{\text{fs}} = 2u_{\text{p}} = U_{\text{imp}}, \tag{9.22}$$

where U_{imp} is the velocity of the impactor, u_{fs} is the free surface velocity and u_{p} is the particle velocity inside the target.

It is also seen from the interferogramm, that the amplitude of the fringe signal is decreased in the middle of the plastic front of the compressive pulse (line *NN*') and is restored to the maximum value at the plateau of the pulse. The contrast of the fringe signal in the middle of the plastic front is $K = 0.727$. In accordance with Eq. (9.20), the velocity variance equals:

$$D_{\max} = \frac{U_{\text{int}}}{\pi}\sqrt{-2\ln K} = \frac{100}{\pi}\sqrt{-2\ln K}(-2\ln K)$$

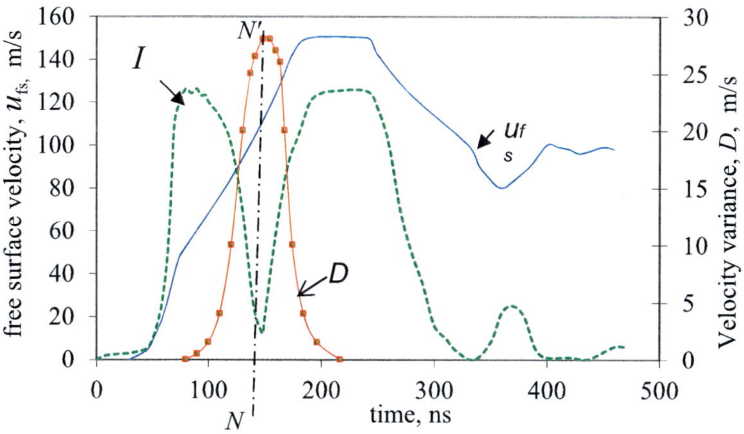

Fig. 9.4 Frequency fringe signal (1), deciphered time-resolved free surface velocity profile, u_{fs}, and velocity variance, *D*, for the loading of a 5 mm steel target at an impact velocity of 150 m/s

$$D_{\max} = \frac{100}{\pi}\sqrt{-2\ln 0,727} = 21.2\,\text{m/s}.$$

Here, the velocity variance $D_{\max}$ corresponds to the maximum width in the middle of the plastic front. At the beginning of the plateau, the amplitude of the fringe signal is restored, which means that the particle velocity distribution relaxes to a δ-function with the velocity dispersion $D^2 = 0$.

The example resented in Fig. 9.4 reflects the situation when the particle velocity distribution is of maximum value in the middle of the plastic front of the compressive pulse. Such microstructure kinetics corresponds to steady plastic waves. If at the moment of transition to the plateau of the compressive pulse, the plastic front has not had time to achieve the steady state, the amplitude of the fringe signal continues to decrease up to the plateau. This, in turn, means that the velocity dispersion increases along the plastic front. A typical free surface velocity history for an unsteady plastic wave is provided in Fig. 9.5. The interference signal and velocity variance for the shock loading of a 5 mm 30CrNi4Mo steel target are also plotted. The fringe signal has not time to recover by the end of the plastic front. The contrast of the fringe signal is seen decreasing, and just before the plateau of the pulse, the velocity variance equals 34 m/s. This means that the plastic wave has not had time to reach an equilibrium state.

In accordance with Eq. (9.20), the contrast of the fringe signal is decreased by e times under the following condition:

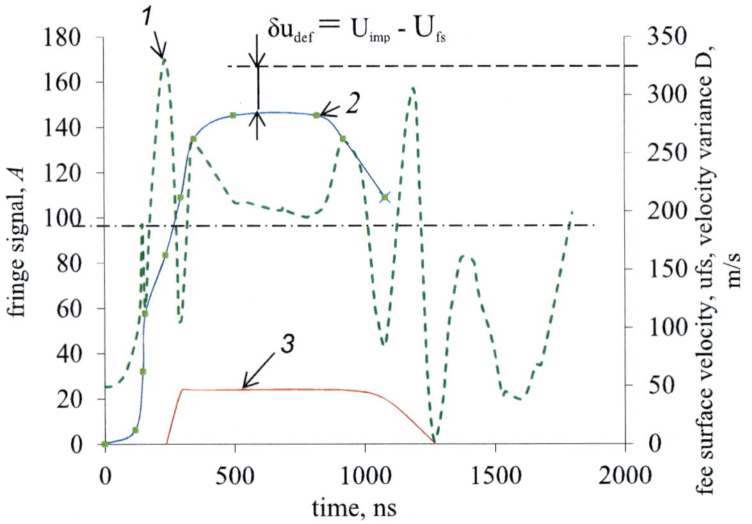

Fig. 9.5 Interference signal (1), time-resolved velocity profile (2) and velocity variance (3) for the shock loading of a 5 mm 38CrNi4Mo steel target at an impact velocity of 320 m/s

$$D = \frac{\sqrt{2}}{\pi} U_{\text{int}}. \tag{9.23}$$

The optimum length of the delay shoulder is determined as follows:

$$l_{\text{d}} = 2n\, l_{\text{r}}, \quad n = 1, 2, 3\ldots, \tag{9.24}$$

where l_{r} is the length of the laser resonator. At the same time, a high value of contrast can be reached in the case when the delay shoulder is much smaller than the length of the laser resonator:

$$l_{\text{r}} \gg l_{\text{d}} \tag{9.25}$$

Typically, the fringe signal can be written as follows:

$$I(t) = I_{\text{o}}\left[1 + 2\sin\left(2\pi \frac{u_{\text{fs}}(t)}{U_{\text{int}}} + \delta\right)\right] \tag{9.26}$$

where $I(t)$ is the current value of the fringe signal, I_{o} is the maximum amplitude of the fringe signal and δ is the initial phase. The current value of the free surface velocity can then be written in the form:

$$u_{\text{fs}} = \frac{U_{\text{int}}}{2\pi}\left[\arcsin\left(\frac{I - I_0}{I_0}\right) - \delta\right] \tag{9.27}$$

Measuring the free surface velocity with the velocity interferometer shows that different materials have different velocity dispersions.

9.5 A Two-Channel Velocity Interferometer

Although application of a short delay shoulder improves the contrast of the fringe signal and allows the mean particle velocity to be measured even for high impact velocities, measurement of the velocity variance is not possible if a single-channel interferometer is used. The single-channel interferometer allows the measuring of the fringe signal contrast only for the points of maximum and minimum values of the sine signal.

Application of a two-channel interferometer with the signals shifted by 90° avoids this difficulty and provides a correct measurement of the velocity dispersion at any point of the velocity profile. The principal schematic of a two-channel velocity interferometer is shown in Fig. 9.6. A two-frequency laser generates two beams, the light polarisation of which is shifted by 90°. After focusing with lens (1), the beams hit

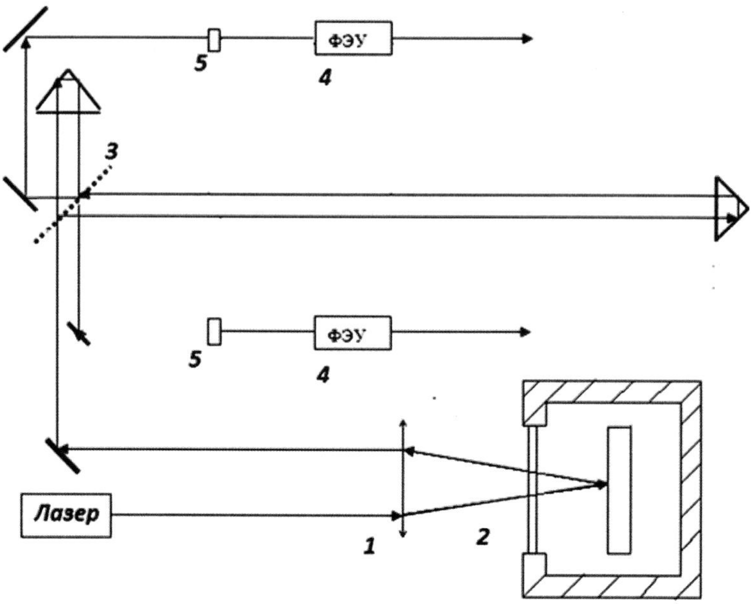

Fig. 9.6 Schematic of the two-channel velocity interferometer

the free surface of the target. The beams reflected from the target land on semi-transparent mirror (3) and pass through the delay shoulders of different lengths. The difference in shoulder paths determines the summary light delay of the interferometer and defines the value of the interference constant, U_{int}. Shifted by 90°, the output signals correspond to different interferometer shoulders, $\boldsymbol{A}_1$ and $\boldsymbol{A}_2$. Each signal contains both polarisations of the laser radiation. Dividing the output signals by the horizontal and vertical components happens at the polaroid (5). As a result, at the output of the photomultipliers, there are two signals shifted in phase by 90°.

In order to demonstrate the possibilities of the two-channel interferometer, let us briefly describe the method used for deciphering the interference signals. Analytically, the signals can be written in the following form:

$$A_1(t) = A_{01}[1 + K(t)\cos(2\pi \frac{u(t)}{U_{\text{int}}} - - - - - - + j)], \tag{9.28}$$

$$A_2(t) = A_{02}[1 + K(t)\sin(2\pi \frac{u(t)}{U_{\text{int}}} - - - - - - - + j + Dj)] \tag{9.29}$$

Here, $A_1(t)$ and $A_2(t)$ are the current values of the fringe signals at the first and second shoulders of the interferometer, A_{01} and A_{02} are the amplitudes of the signals, $K(t)$ is the contrast, φ is the initial, $\Delta\varphi$ is the phase shift between the different shoulders, x, $u(t)$ is the current value of the mean free surface, and U_{int} is the interferometer constant. Let us introduce the following designations:

$$\begin{aligned} a_1(t) &= [A_1(t)/A_{01}] - 1, \\ a_2(t) &= [A_2(t)/A_{02}] - 1, \\ a(t) &= [2\pi u(t)/U_{\text{int}} + \varphi] \end{aligned} \tag{9.30}$$

Taking into account that for the two-channel $\Delta\varphi = \pi/2$, Eq. (3.32) can be written in the form:

$$\begin{aligned} a_1(t) &= K(t)\cos\alpha(t), \\ a_2(t) &= -K(t)\sin\alpha(t) \end{aligned} \tag{9.31}$$

This is the system of two equations with two unknowns, $K(t)$ and $\alpha(t)$. The solution to the system of equations is:

$$\begin{aligned} \alpha(t) &= \text{arctg}[-a_2(t)/a_1(t)] \\ K(t) &= a_1(t)/\cos[\text{arctg}(-a_2(t)/a_1(t)] \end{aligned} \tag{9.32}$$

The mean particle velocity $u(t)$ and the contrast $K(t)$ can be written in the following form:

$$\begin{aligned} u(t) &= \left[\text{arctg}\left(\frac{a_1(t)}{a_2(t)} - \varphi\right)\right] \\ K(t) &= a_1(t)/\cos\left[\text{arctg}\left(\frac{a_2(t)}{a_1(t)}\right)\right]. \end{aligned} \tag{9.33}$$

As $a_2(t)/a_1(t) \to \infty$ at $a_1(t) \to 0$, there is an uncertainty in the determination of $u(t)$ and $K(t)$. In this case,

$$a_1(t) = \frac{\pi}{2} + n\pi \quad \text{where } n = 0, 1, 2, 3 \ldots$$

which yields:

$$\begin{aligned} K(t) &= a_2(t), \\ u(t) &= (U_{\text{int}}/2\pi)\alpha(t) \end{aligned} \tag{9.34}$$

Then, the velocity variance can be found from the equation:

$$D(t) = \frac{U_{\text{int}}}{\pi}\sqrt{-2K(t)} \tag{9.35}$$

As $K(t)$ in this equation is the current value of the contrast, the velocity dispersion is also the current value, so the precision of its determination does not depend on the number of fringe signal beats as in the single-channel interferometer. This is the main advantage of the two-channel interferometer.

It has previously been shown that the velocity dispersion can be determined from the contrast of the fringe signal only under the condition:

$$D < \frac{\sqrt{2}}{\pi} U_{\text{int}}. \tag{9.36}$$

When the dispersion achieves the critical value, the fringe signal disappears, so the interference contrast cannot be determined. The only constraint that can be determined is that the velocity dispersion is greater than $2U_{\text{инт}}/\pi$. A similar situation occurs when the particle velocity distribution suddenly broadens, for example due to a change in the internal kinematic mechanism and the scale of dynamic deformation.

9.6 The Asymmetry of the Mesoparticle Velocity Distribution Function

The equilibrium mesoparticle velocity distribution can be characterised by the two statistical moments of the velocity distribution function: the mean particle velocity and the velocity dispersion. The interference technique allows the measurement of both statistical moments in real time. If the velocity distribution function is not at equilibrium, additional features of the interference technique can be used to measure the degree of asymmetry of the distribution function. It will be shown below that measuring the asymmetry of the velocity distribution function is closely related to determining the shock-induced velocity defect (decrease) of the mean particle velocity.

Typically, the velocity interferometer can be presented in the form of two interacting mutually coherent point light sources with a time delay of radiation of $\tau_d = S_1 S_2 / c$. Here, $S_1 S_2$ is the distance between the point light sources and c is the speed of light (see Fig. 9.7). According to the principles of interference, the electromagnetic waves from the sources S_1 and S_2 arrive at each point of maximum illumination with identical phases which can be written in the form:

$$r_1 - r_2 = m, \text{ where } m = 1, 2, 3 \ldots \tag{9.37}$$

Here, r_1 and r_2 are the radial vectors from the sources S_1 and S_2 which characterise the distance to some arbitrary point M on the surface of maximum illumination. The total number of such surfaces between these sources equals:

$$N = \frac{S_1 S_2}{\lambda}, \tag{9.38}$$

where λ is the wavelength of the light reflected from the target. If the target is at rest, the number of maximum illumination surfaces equals:

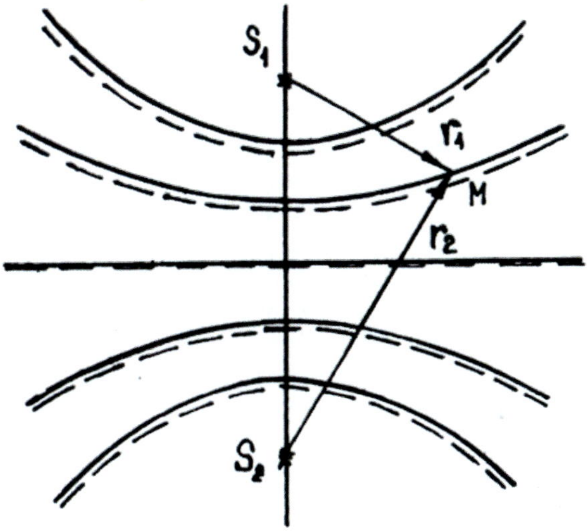

Fig. 9.7 Presentation of direct and delayed beams in the form of two mutually coherent sources

$$N = \frac{S_1 S_2}{\lambda_0}, \tag{9.39}$$

where λ is the wavelength of the light before it interacts with the target. Let $u(t)$ be the current value of the mean free surface velocity. Then, the current wavelength of light reflected from the moving target, in accordance with the Doppler effect, equals:

$$\lambda(t) = \lambda_0 \left[1 + \frac{2u(t)}{c} \right]. \tag{9.40}$$

If the target is at rest, the interference pattern remains static. With the initiation of the motion of the target, the wavelength of the laser radiation reflected from the target begins to change and, according to Eq. (9.38), the number of fringes begins to change as well. It can be calculated how many surfaces of maximum illumination pass through the fixed point in space owing to the change in $\lambda(t)$. This value obviously equals the number of beats of the fringe signal at the output of the interferometer:

$$n = \frac{1}{2}(N - N_0) = S_1 S_2 \left(\frac{1}{\lambda} - \frac{1}{\lambda_0} \right). \tag{9.41}$$

Taking into account Eq. (3.40), one obtains:

$$n = \frac{2\tau_d}{\lambda} u(t), \tag{9.42}$$

which coincides with Eq. (8.10) obtained by Barker [7].

Now, we let the sources S_1 and S_2 radiate the wavelength of light which chaotically changes within the interval $\Delta\lambda = \lambda_{max} - \lambda_{min}$. Applied to velocity interferometer, this means that both the direct and delayed beams interact with the target, the particles of which are distributed in velocity space within the interval $\Delta u = u_{max} - u_{min}$. Again, the direct and delayed laser beams can be presented as a totality of elementary beams, each of which interacts with the concrete particle at the free surface. As a result, the elementary laser beams acquire different Doppler shears, which results in the loss of mono-chromaticity of laser radiation for both direct and delayed laser beams, that is a widening of the radiation spectrum. This radiation can be presented as the radiation of a point source, the wavelength of which chaotically changes within the interval $\Delta\lambda = \lambda_{max} - \lambda_{min}$. The interference of the direct and reflected laser beams at the upper boundary of the spectrum forms the surface of maximum illumination, whereas the interference at the lower boundary of the spectrum results in the minimum illumination (indicated with dotted lines in Fig. 9.7). In accordance with Eq. (3.40), the spatial positions of these surfaces do not coincide. At the beginning, an increase in $\Delta\lambda$ leads to the displacement of the interference pattern with the simultaneous decrease of the contrast followed by the sudden decrease of the number of fringe signal beats. The change of fringe signal beats is determined by the following expression [14]:

$$\Delta N = \frac{2S_1 S_2(\lambda_{max} - \lambda_{min})}{\lambda_{max}\lambda_{min}} = \frac{2S_1 S_2 \Delta\lambda}{\lambda} \tag{9.43}$$

If the free surface velocities are distributed within some interval $\Delta u = u_{max} - u_{min}$, the change of fringe beat number can be obtained as follows:

$$\Delta n = \frac{n}{n-1}\lambda\frac{dn}{d\lambda}\Delta u. \tag{9.44}$$

This formula relates that decrease of fringe beat number with the decrease of the mean value of the particle velocity and can be used for the quantitative characteristic for the defect of particle velocity. As indicated before, in the case of a symmetrical collision between the target and the impactor, the free surface velocity equals the velocity of the impactor ($U_{imp} = U_{fs}$). The difference between these parameters corresponds to the velocity defect, which can be determined by means of measuring the loss in fringe signal frequency beat numbers at the free surface of the target.

9.7 The Determination of the Velocity Distribution at Mesoscale 2 by Using a Line Imaging Velocity Interferometer (LIV) and a Multi-Point VISAR Interferometer

Described above, the point measurements of the velocity dispersion at mesoscale 1 provide data on the particle velocity distribution at this scale (1–10 μm). This is insufficient for development of multiscale models of the shock wave processes. The development of multiscale models of the shock wave phenomena in heterogeneous media requires experimental data on several scale levels. Specifically, one of the challenges in this research area is the in situ determination of the particle velocity distribution at mesoscale 2 (50–500 μm). The point measurements of the velocity dispersion at mesoscale 1 provide information on the velocity distribution only inside the laser spot (60–100 μm), that is, for a single structural element of mesoscale 2, which is insufficient for the information on the response of the target material on impact necessary for multiscale modelling of the shock wave processes.

The first measurements of the velocity distribution at mesoscale 2 have been performed using the line imaging velocity (LIV) interferometer [15] based on the ORVIS technique [15–17]. In this technique, instead of a point laser illumination of the free surface of the target, the laser beam scans the target surface. The linear interferometer produces several fringe signals, each of which corresponds to a different point at the free surface of the target. This technique provides a spatial resolution of about 20 μm for a 200 μm long line and 40 μm for a 2 mm line. All the measured free surface velocity profiles are of different amplitudes, which reflect the presence of the velocity distribution at mesoscale 2. Further improvement of this technique promises to provide a spatial resolution of about 10–20 μm. The in situ measurement of the particle velocity distribution at the mesoscale 1 and mesoscale 2 allows a relationship between the different scales to be established and to predict the multiscale response to shock loading, which, in turn, provides the necessary data for the simulation of multiscale deformation processes.

Another technique for measuring the spatial velocity distribution at mesoscale 2 has been developed in [18]. In this technique, the response of the target to shock loading was measured with seven separate fibre gauges mounted at the free surface of the target. Each gauge was joined to the VISAR interferometer, and its data were recorded with a separate oscilloscope. The closest distance between gauges was 16 μm. The gauge responses provide information on the free surface velocity over the target and allow the velocity distribution at the mesoscale 2 to be inferred. Such distances between gauges excludes the mutual interaction between neighbouring points.

In our approach, the information on the mesoscale 2 velocity distribution is obtained from a series of shots of identical targets and impactor pairs with different impact velocities. This allows the dependence of mesoscale 2 on the strain rate to be revealed and the determination of the optimum strain rate for the highest dynamic strength of the material. In Fig. 9.8, the dependence of the average value of the velocity defect on the impact velocity $\delta u_{\mathrm{def}} = f(U_{\mathrm{imp}})$, together with the deviations from the mean value of the velocity defect, are presented over a series of tests. The region of deviations is restricted by the dotted lines. The deviations are accepted as the velocity variance at mesoscale 2, D_{ms2}.

In this figure, the region of changing the velocity defect, which determines the velocity variance at the mesoscale 2, is defined by the dotted lines.

Fig. 9.8 Dependence of the velocity defect on the impact velocity for **a** the first set and **b** the second set (2) of 40CrNiMo steel

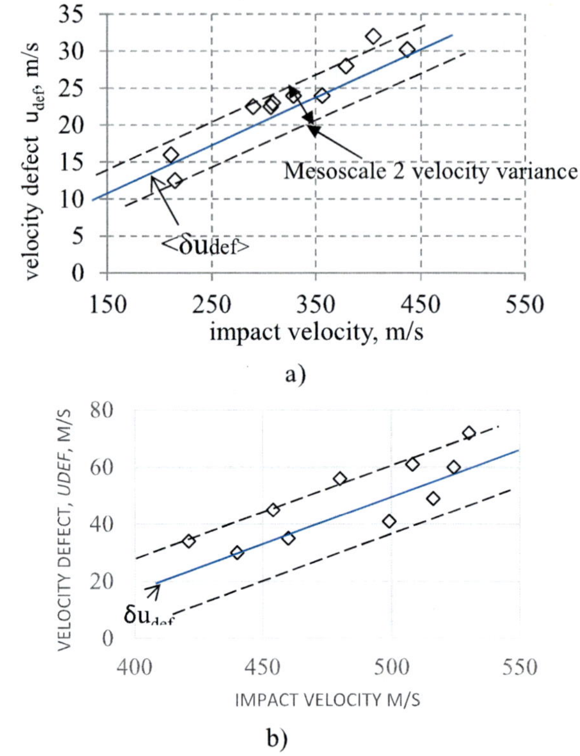

9.8 The Specific Features of the Diagnostic Technique Used

In the case of dynamic straining, the real physical carriers of deformation (point defects, dislocations, twins) that contribute to the shock deformation cannot be identified experimentally. The only value that can be estimated using the interference technique is the spatial scale of the correlated motion of the ECD. In this case, the regions of correlated motion of the ECD are the mesoparticles. The scale of mesoparticle $\boldsymbol{d}$ is estimated to lie within the interval $\lambda_0 \ll \boldsymbol{d} \ll L$, where λ_0 is the wavelength of laser radiation and L is the diameter of the laser beam. For $\lambda_0 \sim 0.6$ μm (He–Ne laser) and a laser spot dimension of ~60 μm, the mean size of the mesoparticle equals approximately 6 μm, which is much smaller than the grain size for most crystalline solids, but corresponds to mean dimension of the elementary carriers at mesoscale 1.

The specifics of the diagnostics used are that the laser spot of the interferometer (~60 μm) coincides with the dimension of the single ECD at mesoscale 2 (50–500 μm). This means that the parameters measured for shock waves: the free surface velocity profile $u(t)$, the velocity variance D and the velocity defect δu_{def} are related to a single mesoscale 2 particle. In this situation, during the shock loading the velocity interferometer records: (i) the mean velocity of a single mesoparticle as a whole, (ii) the velocity losses related to the motion of a single mesoparticle and (iii) the velocity distribution in the background of the single mesoscale 2 mesoparticle. For example, the micrographs of relict mesoparticles in the copper spall zone shown in Fig. 4.3 of Chap. 4 display a family of shear bands inside the mesoscale 2 mesoparticle. The spatial scale of the mesostructure inside the mesoparticle corresponds to mesoscale 1, since the mean distance between the shear bands equals 5–10 μm. Thus, it may be concluded that the velocity distribution measured with the interferometer in copper related to the stochastic behaviour of shear bands inside this mesoparticle alone.

Depending on the type of the material, the dynamic deformation can flow in the form of translational or rotational motions of mesoparticles. In the first case, the interferometer registers the free surface velocity profile $u_{fs}(t)$ corresponding to the translational motion of the mesoparticle as a whole. When the diagnostic technique concerns a single element of mesoscale 2, the fringe signal is sensitive to the type of motion of the mesoscale 2 structure—rotational or translational. The typical interference signals for both cases of material response are provided in Fig. 9.9. The first signal reflects the response of a 1565 aluminium alloy, whereas the second signal is obtained for an 1561 aluminium alloy. According to the working principles of the interferometer, the irreversible displacement of the fringe signal to the upper level of the interference fringe pattern is due to a loss of intensity of the laser beam reflected from the free surface of target (see Fig. 9.9b). This occurs in the case when the ECD at the mesoscale 2 rotates as a whole. In this case, the laser beam deviates so the fringe signal cannot be registered. Microstructural investigations show that the first material presents a highly textured grain structure, which suppresses the rotational motion of grains, whereas the second material presents a viscous medium, the grains of which allow the rotational motions of mesoparticles.

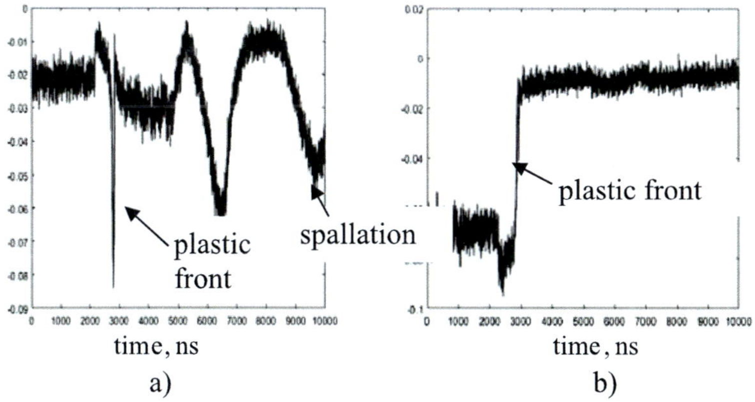

Fig. 9.9 Fringe signals for **a** translational and **b** rotational motion of a deformed medium

In the first case, the full free surface profile, including the elastic precursor, the plastic front and the back front is measured; whereas in the second case, only the elastic precursor and the plastic front are registered.

References

1. Meshcheryakov YuI, Prokuratova EI (1995) Kinetic theory of continuously distributed dislocations. Int J Solids Struct 32:1711–1726
2. Prokuratova EI, Indeitzev DI (1995) The conditions for the dislocation distribution localization under dynamic loading. Dymat J 6(¾):229–233
3. Meyers MA, Carvalho MS (1976) Shock-front irregularities in polycrystalline metals. Mate Sci Eng 24:131
4. Case S, Horie Y (2007) Modeling the shock response of polycrystals at the mesoscale. J Mech Phys Solids 35:589–614
5. Wallace DC (1980) Irreversible thermodynamics of flow in solids. Phys Rev B 22(4):1477
6. Barker LM, Hollenbuch RE (1965) Interferometric technique for measuring the dynamic mechanic properties of materials. Rev Sci Instrum 36(11):1617–1621
7. Barker L.M (1967) Fine structure of compressive and release wave shapes in aluminum measured by the velocity interferometer technique. In: Symposium on high dynamic pressure. Paris, pp 369–382
8. Clifton RI (1970) An analysis of the laser velocity interferometer. J Appl Phys 41(12):5335–5337
9. Zlatin NA, Mochalov SM, Pugachev GS, Bragov AM (1981) Laser differential interferometer. J Tech Phys 51(7):1503–1506 (Russian edition)
10. Barker LM, Hollenbach RE (1972) Laser interferometer for measuring high velocities of any reflecting surfaces. J Appl Phys 43(6):4669–4679
11. Asay JR, Barker LM (1974) Interferometric measurement of shock-induced particle velocity and spatial variation of particle velocity. J Appl Phys 45(5):2540–2550
12. Meshcheryakov YuI, Divakov AK (1994) Multiscale kinetics of structure and strain-rate dependence of the materials. Dymat J 1(4):271–287
13. Cohran S, Banner D (1977) Spall studies in uranium. J Appl Phys 48(7):2729–2937
14. Baiborodin YV, Kriksunov LZ, Litvinenko ON (1978) Handbook on laser Tekhnika. Kiew. Tekhnika, p 288 (Russian edition)
15. Chhabildas LC, Trott WM, Reinhart WD, Cogar JR, Man GA (2001) Incipient spall studied in tantalum-microstructural effects. In: Workshop on shock dynamics and non-equilibrium mesoscopic fluctuations in solids. 23 June 2001. Atlanta GA, USA
16. Trott WM, Castaneda JN, O'Hare JJ, Baer MR, Chhabildas LC, Knudson MD, Davis JP, Asay JR (1995) Examination of the mesoscopic scale response of shock compressed heterogeneous materials using a line-imaging velocity interferometer. In: Staudhammer KP, Murr LE, Meyers MA (eds) Shock-wave and high-strain-rate phenomena, pp 647–654
17. Asay JR, Chhabildas JC (2002) Paradigms and challenges in shock wave research. In: Horie Y, Davison L, Thadhani NN (eds) High-pressure shock compression of solids. Old paradigms and new challenges. Springer, NY, pp 56–119
18. Barker LM (2000) Multi-beam VISARS for simultaneous velocity v/s time measurements. In: Furnish MD, Chhabildas LC, Nixson RS (eds) Shock compression of condensed matter-1999. AIP conference proceedings, p 1–56396–923–2–00
19. Baumung K, Singer J, Rasorenov SV, Utkin AV (1995) In: Schmidt SC, Tao WC (eds) Shock compression of condensed matter-1995. AIP conference proceedings. vol 370, pp 1015–1918

Chapter 10
On the Kinetic Nature of Structural Instability and Localisation of Dynamic Deformation

Abstract Two types of mechanism for the structural instability are studied. The first mechanism supposes the development of structural instability in a threshold manner at the critical strain rate. To initiate that mechanism, the shock loading of material is carried out with gradually increased impact velocities. In this case, under certain strain rates, the initially uniform and quasi-equilibrium dynamic deformation transitions into heterogeneous deformation. Thus, the first type of instability is initiated in accordance with Scenario 1: quasi-equilibrium dynamic deformation below critical strain rate → threshold of instability at the critical strain rate → non-equilibrium dynamic deformation above critical strain rates. The second type of instability flows in accordance with the opposite scenario, Scenario 2: shock-induced non-equilibrium dynamic deformation → transition to quasi-equilibrium dynamic deformation. The second type of structure is nucleated at small spatio-temporal scales under high non-equilibrium conditions. In our study, in order to investigate the second type of instability, shock loading of thin specimens has been performed. The nanosecond stress pulses were initiated in two ways: (i) in-plane collision experiments and (ii) using a high-power electron pulse beam. A scenario for developing the kinetic mechanism of structural instability includes the following stages.

1. The formation of the chaotically distributed dislocation pile-ups.
2. At the critical strain rate, the velocity distribution arrives at the non-equilibrium stage when the mesoscale velocity defect emerges. The velocity defect appears when the rate of change of the velocity dispersion at the mesoscale becomes equal to or higher than the rate of change of the mean particle velocity.
3. The critical value of the velocity defect is the starting point for: (i) non-reversible momentum exchange between the mesoscale and the macroscale, (ii) the formation of mesoscale 2 structures and (iii) the motion of mesoscale 2 structural element as a whole. The motions occur independently at numerous points of the shock-deformed target, which indicates the transition of the material unto structurally unstable state.

Y. Meshcheryakov, *Multiscale Mechanics of Shock Wave Processes*, Shock Wave and High Pressure Phenomena, https://doi.org/10.1007/978-981-16-4530-3_10

10.1 Introduction

Two types of mechanism for the structural instability are studied. The first mechanism supposes the development of structural instability in a threshold manner at the critical strain rate. To initiate that mechanism, the shock loading of material is carried out with gradually increased impact velocities. In this case, under certain strain rates, the initially uniform and quasi-equilibrium dynamic deformation transitions into heterogeneous deformation. Thus, the first type of instability is initiated in accordance with Scenario 1: quasi-equilibrium dynamic deformation below critical strain rate → threshold of instability at the critical strain rate → non-equilibrium dynamic deformation above critical strain rates.

The second type of instability flows in accordance with the opposite scenario, Scenario 2: shock-induced non-equilibrium dynamic deformation → transition to quasi-equilibrium dynamic deformation. The second type of structure is nucleated at small spatio-temporal scales under high non-equilibrium conditions. In our study, in order to investigate the second type of instability, shock loading of thin specimens has been performed. The nanosecond stress pulses were initiated in two ways: (i) in-plane collision experiments and (ii) using a high-power electron pulse beam (Chap. 8).

10.2 Interscale Momentum Exchange and the Kinetic Criterion for Transition into a Structurally Unstable State

According to Duvall's classification [1], there are three types of shock-wave decay in solids: (i) geometric decay, (ii) hydrodynamic decay and (iii) Maxwell decay. The first type of decay is related to the spatial geometry of the shock wave front. For spherical waves, the decay is proportional to $1/r^2$; for cylindrical waves, the decay is proportional to $1/r$; and for planar waves, the decay does not depend on distance at all. Hydrodynamic decay occurs because the back front of the compressive pulse propagates with a higher velocity than the first front, which leads to cutting off the top of the stress pulse. The Maxwell decay is a sequence of dissipative processes inherent to plastic deformation.

In the case of the dynamic deformation of heterogeneous media, an additional type of shock wave decay can occur. This is the "fluctuative" decay, which is peculiar to non-equilibrium processes in deformed media. During the transition from homogeneous to heterogeneous flow, the redistribution of momentum occurs, which is accompanied by the particle velocity distribution, the typical kinetic effect. The decay of the shock wave in a heterogeneous medium can be considered as an example of the interaction between the noise (here, the noise is the mesoparticle velocity fluctuations) and the mean particle velocity. It has been shown in Chap. 5 that unlike quasi-equilibrium processes, where the noise-induced transitions can be initiated at the critical intensity of the noise [2], in the case of non-equilibrium processes, the important role in structural transition is played not only by the intensity of the noise (velocity dispersion) but the rate of change of the noise intensity as well.

Our approach is based on four hypotheses, which follow from theoretical and experimental investigations of the dynamic processes in solids.

The propagation of a shock wave in a heterogeneous medium is considered to be a superposition of two modes: (i) the propagation of an approximately rectilinear front with mean particle velocity $u(x, t)$ and (ii) the random motions of the medium at the mesoscale around the mean velocity as result of the fluctuative strain fields induced by the random internal stresses.

This hypothesis has been used in [3] to describe the meso–macro momentum exchange in a dynamically deformed solid.

1. The intensity of the meso–macro momentum exchange, in the form of the velocity defect, δu_{def}, has been determined by the rate of change of the mesoparticle velocity dispersion, D^2:

$$\delta u_{\mathrm{def}} = \frac{\mathrm{d}D^2}{\mathrm{d}u}. \tag{10.1}$$

2. The random motions of the elementary carriers of deformation at the mesoscale are quantitatively characterised by the mesoparticle velocity distribution, the intensity of which (in the form of the mesoparticle velocity variance, D) is accepted to be proportional to the strain rate [4]:

$$D = R\frac{\mathrm{d}\varepsilon}{\mathrm{d}t}. \tag{10.2}$$

   A similar dependence is known to exist in turbulence, where the intensity of the turbulent pulsations is proportional to the particle acceleration [5].
3. In the case of non-equilibrium deformation processes, the mesoscale is divided into two sublevels—mesoscale 1 (1–10 μm) and mesoscale 2 (50–500 μm). Both types of mesostructure exist only during the non-equilibrium stage of dynamic deformation. Elementary carriers of dynamic deformation at the mesoscale 1 are thought to be short-lived (150–200 ns) single-sign dislocation pile-ups. The lifetime of mesoscale 1 structures is of the order of 150–200 ns [6–8]. The mesoparticles are assumed to experience two types of interaction: (i) long-range mutual interactions and (ii) interactions with the medium.
4. The mesoscale 2 structures are nucleated during the transition of the medium into a structurally unstable state.

In the present investigation, the processes of dynamic deformation are considered from the perspective of the physical kinetics of the mesoscale medium [8]. Often, the random motions of mesoparticles can be characterised by the PVDF $f(r, v, t)$. In the case of a quasi-equilibrium situation, the mean particle velocity, $u(x, t)$, and the particle velocity dispersion, $D^2(x, t)$, are the first and the second statistical moments of the PVDF.

$$u(x,t) = \int v f(x,v,t)\mathrm{d}v;$$
$$D^2(x,t) \int (v-u)^2 f(x,v,t)\mathrm{d}v, \tag{10.3}$$

As distinct from the dislocations, the temporal behaviour of the above characteristics can be measured in real time using an interference technique. The main characteristic that provides quantitative data on the momentum exchange between scales is the defect of particle velocity, δu_{def}. In the case of non-equilibrium systems, such as a dynamically deformed medium, the mean particle velocity, $u(t)$, the particle velocity dispersion, $D^2(t)$ and the velocity defect, δu_{def}, are not independent values [3]:

$$\delta u_{\mathrm{def}}(x,t) = u(x_0,t) - u(x,t) = \frac{1}{2}\frac{d(D^2)}{\mathrm{d}u}. \tag{10.4}$$

Here, $u(x,t)$ is the current value of the mean particle velocity; $u(x_0,t)$ is a reference mean particle velocity value; and $\delta u_{\mathrm{def}}(x,t)$ is the current value of the velocity defect, which determines the change in the mean particle velocity resulting from the interscale momentum exchange. The right-hand side of Eq. (10.4) characterises the intensity of the interscale momentum exchange, which equals the rate of change of the particle velocity dispersion. Physically, Eq. (10.4) determines the change of the mean particle velocity due to the changing intensity of the velocity fluctuations at the mesoscale. Equation (10.4) reflects the current character of the interscale momentum exchange. Averaging the interscale momentum exchange yields:

$$\frac{1}{\tau}\int\limits_{t-\tau_i}^{t}\left(\frac{1}{2}\frac{\mathrm{d}D^2}{\mathrm{d}u}\right)\mathrm{d}s \le \delta u_{\mathrm{def}}^{\mathrm{cr}}, \tag{10.5}$$

where τ is the minimum time required for a particle velocity defect to be formed. The time τ can be considered to be the "incubation time" for interscale momentum exchange. Then, $\delta u_{\mathrm{def}}^{\mathrm{cr}}$ is the threshold value of the velocity defect, corresponding to the transition of a solid into a structurally unstable state. In this case, Eq. (10.5) can be considered as the criterion for the transition of a solid into a structurally unstable state. Then, Eq. (10.5) can be written in the form:

$$\frac{1}{\tau}\int\limits_{t-\tau_i}^{t}\left(D\frac{\mathrm{d}D}{\mathrm{d}t}\Big/\frac{\mathrm{d}u}{\mathrm{d}t}\right)\mathrm{d}s \le \delta u_{\mathrm{def}}^{\mathrm{cr}} \tag{10.6}$$

The incubation time is the time for the mesoparticle velocity distribution function to come to equilibrium state (also referred to as the "time of Maxwellisation"). Scant data exist on the time of Maxwellisation of the velocity distribution function. For example, the numerical simulation in [9] provides data on the time of Maxwellisation of the PVDF in copper (11.5 ns). As the time of Maxwellisation depends on the type of material, the dimensions of the mesoscale 2 structural elements are different for different materials.

The velocity defect emerges when the strain rate is higher than the critical value corresponding to the threshold of structural instability. Unlike the heat mechanism of [10, 11], the criterion (10.5) is of kinetic nature. In accordance with that criterion, the initiation of structural instability occurs due to the critical relationship between the rate of change of the velocity variance at the mesoscale and the rate of change of the mean particle velocity:

$$\left(\frac{\mathrm{d}D}{\mathrm{d}t} \Big/ \frac{\mathrm{d}u}{\mathrm{d}t}\right) \leq 1. \tag{10.7}$$

A scenario for developing the kinetic mechanism of structural instability includes the following stages.

1. The formation of the chaotically distributed dislocation pile-ups [6–8]. The long-range interaction of pile-ups results in the mesoparticle velocity distribution.
2. At the critical strain rate, the velocity distribution arrives at the non-equilibrium stage when the mesoscale velocity defect emerges. The velocity defect appears when the rate of change of the velocity dispersion at the mesoscale becomes equal to or higher than the rate of change of the mean particle velocity.
3. The critical value of the velocity defect is the starting point for: (i) non-reversible momentum exchange between the mesoscale and the macroscale, (ii) the formation of mesoscale 2 structures and (iii) the motion of mesoscale 2 structural element as a whole. The motions occur independently at numerous points of the shock-deformed target, which indicates the transition of the material unto structurally unstable state.

The above scenario is appropriate for the non-equilibrium processes at which the mesoparticle velocity distribution function is not symmetric. For the non-equilibrium velocity distribution, the number of particles with velocities smaller the mean particle velocity is greater than the number of particles with velocities that are greater than the mean particle velocity. In this case, the Landau decay occurs, which is registered as decrease, or defect, in the mean velocity. A typical situation is obtained in [9], where the simulation of the shock wave process reveals the non-equilibrium PVDF in shock-deformed copper.

These enumerated stages of development of localised structural instability are confirmed by experiments in which the scenarios were tested with high spatio-temporal resolution equipment (~0.6–1.0 ns). This equipment allows the single act of shock loading to produce data in real time on both the mean motion of a single mesoscale 2 structural element and the velocity dispersion on the background of that structural element (see Chap. 9). Thus, one records the mesoscale 1 velocity dispersion and the velocity defect for the single mesoscale 2 structural element. This freezes the moment of the transition into the structurally unstable state which can be called the kinetic criterion of the shock wave structural instability. The velocity defect begins to increase at the strain rate where the material transits into the structurally unstable state. Below, we present examples of the transitions of shock-deformed material into structurally unstable states.

The velocity defect is sometimes attributed to the interaction of the elastic precursor reflected from the free surface of the target and the running plastic front [12]. However, this hypothesis is not physically justified. The propagation of the elastic precursor is based on the Lennard–Jones or Morse atomic potentials, whereas the motion of the medium in a plastic wave is not based on a potential. Experiments also show that the interaction between plastic front and elastic wave is not registered. Instead, there is a small cusp at the plastic front, which indicates the moment the plastic front and elastic precursor meet, but does not decrease the plastic front amplitude [13]. In [3], the free surface profile for the shock-loaded copper target demonstrates a large decrease in the particle velocity at the plateau of the compressive pulse (172.6 m/s) while the elastic precursor amplitude equals 2–3 m/s. Such a small elastic precursor cannot be responsible for the large value of the velocity defect, even if the interaction between the plastic front and the reflected precursor existed. Conversely, the velocity defect for shock-loaded titanium alloys does not exceed 20–30 m/s, although the elastic precursor is on the order of 170 m/s in experiments of this type [14.]. The large velocity defect appears only at impact velocities higher than the threshold of the structural transition. This is a local effect which is observed using the experimental technique in the present study.

10.3 On the Resonance Excitation of Mesoscale

In the light of the above analysis, three questions arise: (i) what is the physical mechanism responsible for the nucleation of the mesoscale 2, (ii) what shock wave parameters determine the dimensions of the mesoscale 2 structures and (iii) what is the physical mechanism for the trigger-like transition from one scale to another? To answer these questions, let us consider the propagation of a plane shock wave in a relaxing medium. In the case of a steady shock wave, two important relationships have previously been found [3, 4]. The first relates the particle velocity dispersion, D^2, and the velocity defect, δu_{def}, in Eq. (10.1). The second equation relates the velocity variance, D, and the strain rate, (Eq. (10.2)). We use Eqs. (10.1) and (10.2) to describe the response of a relaxing medium to shock loading.

In the case of the one-dimensional propagation of a shock wave, the balance equations for momentum conservation and medium continuity take the form

$$\rho u_t - \sigma_x = 0; \tag{10.8}$$

$$u_x - \varepsilon_t = 0. \tag{10.9}$$

The constitutive equation for a relaxing medium has been introduced by Duvall [15] and Taylor [16] as:

$$\sigma - \rho C_l^2 \varepsilon = -2\mu\varepsilon^p, \tag{10.10}$$

Or, in the differential form

$$\sigma_t - \rho C_l^2 \varepsilon_t = -F. \tag{10.11}$$

Here, C_l is the longitudinal sound velocity and μ is the shear modulus. The relaxation function

$$F = 2\mu \frac{\mathrm{d}\varepsilon^p}{\mathrm{d}t} \tag{10.12}$$

is determined through Orowan's equation for the mean density and mean velocity of dislocations

$$\frac{\mathrm{d}\varepsilon^p}{\mathrm{d}t} = bN_\mathrm{d} V_\mathrm{d}.$$

In the present study, unlike in the dislocation structure, the stress relaxation in the dynamically deformed medium is accepted to be realised only through the motion of ECD at the mesoscale. Dislocations and other carriers of the deformation at atom-dislocation scale form the mesoparticles, whereas the mesoparticles themselves are self-consistent carriers of deformation. In this situation, the stress relaxation is determined only by the velocity defect:

$$\frac{\mathrm{d}\varepsilon^p}{\mathrm{d}t} = \frac{1}{C_\mathrm{l}} \frac{d(\delta u_\mathrm{def})}{\mathrm{d}t}. \tag{10.13}$$

The advantage of our approach is that the relaxation model does not incorporate the parameters of the dislocations that cannot be controlled under conditions of dynamic straining, such as the mean density and the mean velocity of the dislocations. Separate from the dislocation structure, the defect of particle velocity can be registered in real time. In this case, the relaxation function takes the form:

$$F = \frac{2\mu}{C_l} \frac{d(\Delta u_\mathrm{def})}{\mathrm{d}t}. \tag{10.14}$$

The equation system (10.8)–(10.9) can be reduced to a second-order differential equation,

$$\rho\varepsilon_{tt} - \sigma_{xx} = 0. \tag{10.15}$$

Substitution of Eq. (10.1) into Eqs. (10.2) and (10.13) yields:

$$\frac{\mathrm{d}\varepsilon^p}{\mathrm{d}t} = \left(R^2/C_\mathrm{l}^2\right) \frac{d^2\varepsilon}{\mathrm{d}t^2}. \tag{10.16}$$

Then, Eq. (10.15) is reduced to:

$$\varepsilon_{tt} - \rho C_l^2 \varepsilon_{xx} - 2\left(\mu R^2 / \rho C_0^2\right)\varepsilon_{xxtt} = 0. \tag{10.17}$$

In the case of a steady propagation of a shock wave, the single variable $\varsigma = x - C_{\mathrm{p}}t$ can be used. Equation (10.17) thus takes the form:

$$(C_{\mathrm{l}}^2 - C_{\mathrm{p}}^2)\varepsilon_{\varsigma\varsigma} + 2\left(\mu R^2 / \rho C_{\mathrm{l}}^2\right)\varepsilon_{\varsigma\varsigma\varsigma\varsigma} = 0. \tag{10.18}$$

Exchanging $\varepsilon_{\varsigma\varsigma} = \psi$ leads to the equation for an oscillator:

$$(C_{\mathrm{l}}^2 - C_{\mathrm{p}}^2)\psi + 2\left(\mu R^2 / \rho C_{\mathrm{l}}^2\right)\Psi_{\varsigma\varsigma} = 0.$$

The frequency of the oscillations equals

$$\omega = \frac{1}{2}\frac{\rho C_{\mathrm{l}}^2 (C_{\mathrm{l}}^2 - C_{\mathrm{p}}^2)}{\mu R^2}. \tag{10.19}$$

Physically, the oscillatory regime of shock wave propagation is a sequence of positive back coupling between the velocity variance and the velocity defect. The temporal resolution of laser diagnostics (0.6–1.0 ns) allows the recording of the fine structure of the plastic wave, including high-frequency oscillations at the mesoscale. Where observed, the oscillations are excited at the top of the plastic front. Figure 10.1 demonstrates the velocity oscillations in a 7 mm 1565 aluminium alloy target loaded at an impact velocity of 590 m/s. The oscillations are seen at the impact velocity, which is higher than the threshold of the structural transition. The spatial period of oscillations equals ~ 50 μm, which corresponds to a mesoscale 2 structural element.

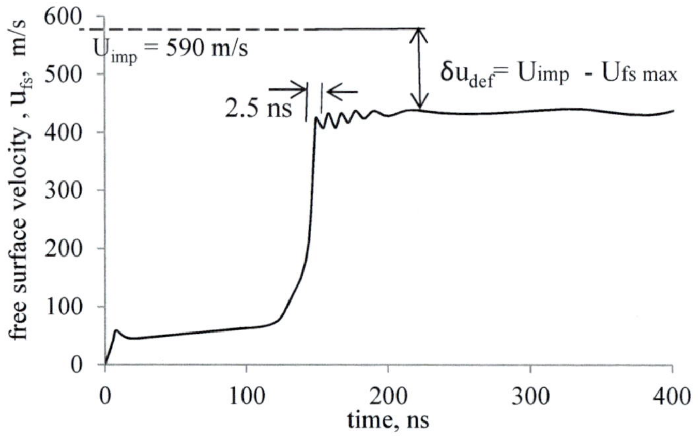

Fig. 10.1 Free surface velocity profile, u_{fs}, for a 7 mm 1565 aluminium alloy target loaded at an impact velocity of 590 m/s

Although oscillations are initiated due to the stochastic features of the dynamically deformed structure in the form of particle velocity pulsations, the transition happens due to swings in the high-frequency oscillations of the mesostructure. In the case of dynamic straining, the meso–macro transition is not a direct transition from the chaotic motion of the structural elements to a translation motion as in a "noise-induced transition" [2]. In a non-equilibrium situation, this transition is thought to be accomplished through the oscillation regime of strain at the mesoscale.

10.4 Scenario 1: Quasi-Equilibrium Dynamic Deformation Below Critical Strain Rate → Non-Equilibrium Dynamic Deformation Transition

One of the first models for structural instability has been developed in [10, 11]. In their approach, the main role is assigned to temperature softening of the material. The authors supposed that the joint action of high hydrostatic pressure and temperature led to an avalanche-like increase of temperature in the regions local to the high-intensity plastic flow, for example, within the local zones of high concentrations of dislocations. This, in turn, resulted in a local decrease of viscosity and a further decrease in the shock front duration. The temperature in this zone begins to increase, since the mechanism of thermo-conductivity has not had time to remove the heat from the warming region, which leads to a further decrease in the viscosity. This model has been used to explain of the nucleation of shear bands initiated by the shock loading of 6061-T6 aluminium alloy [17]. Measuring the temperature increase within a shear band in the dynamic deformation process is not possible in real time. To date, the only characteristic that could be measured in shock-deformed specimens is the mean temperature of the specimen during experiments of [18].

An alternative approach to monitoring shock-induced structural transformation is developed in [19]. The spatial positions of localised formations are determined by the interference of stress waves initiated by the velocity distribution at the mesoscale.

Simulations of the shock-induced structural transitions have been performed in [9, 20]. By using the cellular automata technique, the authors have shown that there is a threshold strain rate when the laminar plastic flow transits into turbulent flow. This simulation reproduces the experiments of [3] and indicates that at high impact velocity, the shock-induced structure in a solid is of a turbulent nature and the transition into turbulent processes happens in the threshold manner.

The approach developed in [21] is based on SEM and TEM investigations of the microstructures of post-shock specimens. The dynamic localisation is speculated to be caused by the interference of tensile stresses reflected from different surfaces. The interference processes of tensile stress waves cannot be traced because the spatio-temporal resolution of the equipment does not allow such high-velocity processes to be monitored in real time. Therefore, both approaches present hypotheses, so the parameters of the localisation process can be considered only as fitting parameters.

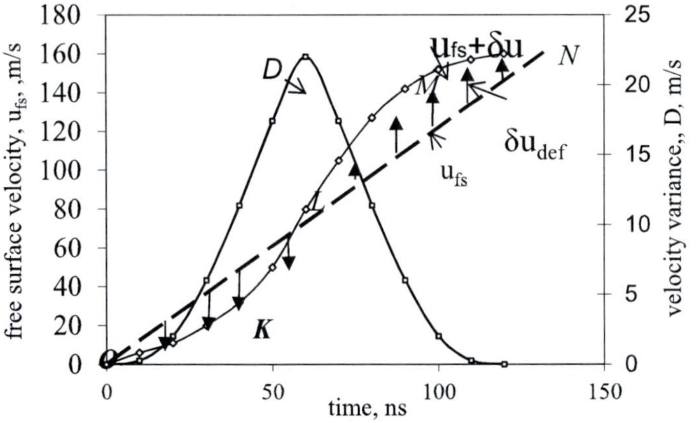

Fig. 10.2 Qualitative picture of the momentum exchange between the mesoscale and the macroscale for a steady shock front

In our research, we developed a new approach based on the real-time observation of the kinetic characteristics of shock wave processes. The mesoparticle velocity dispersion is an important characteristic of dynamic deformation processes in heterogeneous media and is the quantitative measure of the particle velocity scattering. The behaviour of the velocity dispersion is described in the form of Eq. (10.1), which relates the velocity defect and the mesoscale velocity dispersion (see Chap. 5). This relationship determines the value of the velocity defect for the quasi-equilibrium processes of dynamic deformation. The above relationship allows the interactions between the velocity variance, the mean particle velocity and the velocity defect over the steady plastic front to be presented qualitatively, as shown in Fig. 10.2. The dotted line *OLN* gives the position of the mean velocity profile, $\overline{u}_{fs}$, while the curve *OKLMN* is the sum of the mean velocity and the velocity defect, $u_{fs} = \overline{u}_{fs} + \delta u_{def}$ (Fig. 10.2).

In the steady shock front, in accordance with Eq. (10.1) and Fig. 10.2, during the first half of the velocity profile, the velocity defect decreases the mean velocity, whereas during the second part of the profile, the mean velocity is increased by the same value.

In Fig. 10.3, the free surface profile for a 2.2 mm 1565 aluminium alloy target loaded at an impact velocity of 422 m/s is provided. The particle velocity variance is seen to be maximal in the middle of the shock front. The relaxation time of the distribution function to equilibrium state is comparable to the duration of shock front, allowing the evolution of the velocity distribution function to be qualitatively presented, as shown in Fig. 4.7 (Chap. 4). For unsteady shock waves, the mesoparticle velocity distribution exists not only at the front of the compressive pulse but over the total compression pulse, including the plateau and back front (see Fig. 10.4).

As indicated in Chap. 9, when the laser beam is focused to 60–70 μm, the strength characteristics and instability threshold correspond to a single mesoscale 2 element

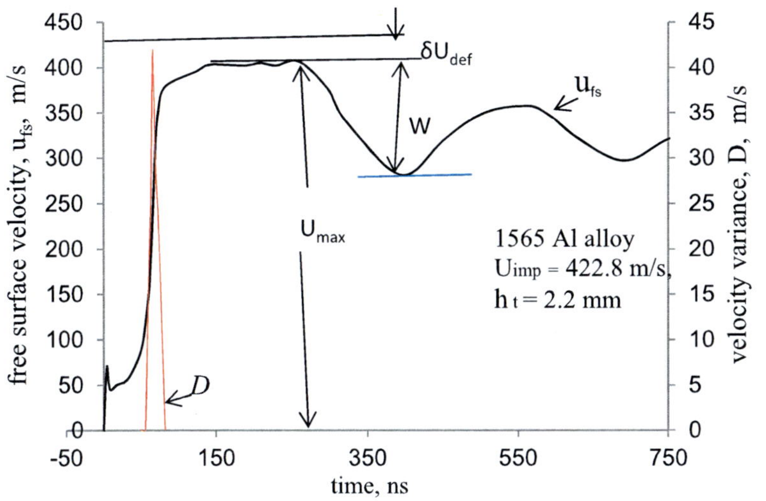

Fig. 10.3 Free surface velocity profile, $u_{fs}(t)$, and velocity variance profile, $D(t)$, for a 1565 aluminium alloy target of 2.2 mm thickness loaded at an impact velocity of 422.8 m/s

(50–500 μm). At the same time, in the background of the single mesoscale 2 structural element, there are a lot of mesoscale 1 (1–10 μm) structural elements. In our experiments, the interferometer registers the velocity distribution of the ECD for mesoscale 1. This is the main advantage of our experimental approach compared to those based on the VISAR technique or the use of different gauges.

The intensity of the macro–meso momentum exchange for both equilibrium and non-equilibrium processes is quantitatively characterised by the velocity defect, δu_{def}. In the case of shock wave experiments, the velocity defect is determined based on independent measurements of the impact velocity, U_{imp}. According to the free surface approximation, in the case of a symmetrical collision of the impactor and the target, the particle velocity equals half the impact velocity, that is $u_p = \frac{1}{2}U_{imp}$. In contrast, when the shock wave arrives at the free surface of the target, the mean velocity doubles. If the losses of velocity are absent during the shock wave propagation through the target, the impact velocity equals the free surface velocity, $U_{imp} = U_{fs\ max}$. Specifically, in the case of the shock loading of iron, this equality has been experimentally confirmed in [22]. If the strain rate is higher than the critical value corresponding to the structural heterogenisation, this condition is not accomplished. In this case, the velocity defect emerges as the difference between the impact velocity and the free surface velocity.

The dependence of the velocity defect on the impact velocity for D16 aluminium alloy is presented in Fig. 10.5. Below an impact velocity of 382 m/s, the velocity defect experiences a weak dependence on the impact velocity. However, at impact velocities higher than $U_{imp} = 382$ m/s, the velocity defect begins to grow much faster, achieving the value of $\delta u_{def} = 150$ m/s at $U_{imp} = 451$ m/s. The sudden increase of mesoscale 2 structural elements begins the fragmentation of the material structure.

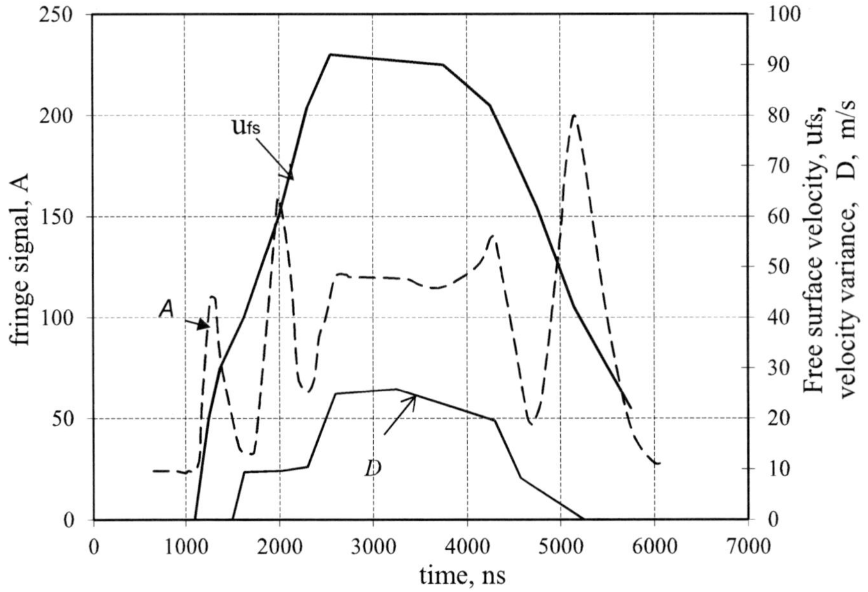

Fig. 10.4 Fringe signal (A), the free surface velocity profile, $u_{fs}(t)$, and the velocity variance profile, $D(t)$, for a 2 mm D16 aluminium alloy target loaded at an impact velocity of 235 m/s

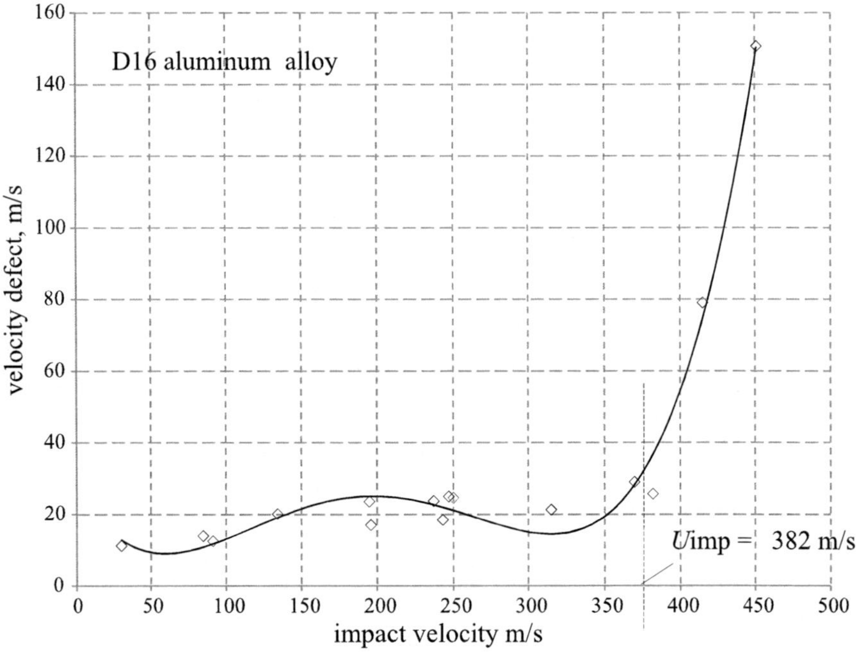

Fig. 10.5 Velocity defect, $\delta u_{fs}(t)$, versus impact velocity for D16 aluminium alloy

The large values of the velocity defect are explained by the specifics of the experimental technique used to measure the free surface velocity. As outlined in previous chapters, the diameter of the laser spot at the free surface of the target is approximately 60–70 μm. This allows the measurement of the response of a single structural element of mesoscale 2. Before the transition of the material into a structurally unstable state, the velocity variances at the mesoscale 1 and mesoscale 2 satisfy the relationships:

$$D_{\mathrm{ms2}} < D_{\mathrm{ms1}} \quad \text{and} \quad D_{\mathrm{ms1}} \simeq D_{\mathrm{ms2}}$$

In this case, the velocity interferometer registers the macroscopic mean particle velocity as the result of averaging the mesoscale velocity fluctuations. In this state, mesoscale 2 is not initiated. If, however, the material is in a structurally unstable state, the opposite relationship is true:

$$D_{\mathrm{ms2}} > D_{\mathrm{ms1}}.$$

The velocity interferometer registers the mean velocity of a single structural element of mesoscale 2. In this case, the local velocity scattering can achieve large values. A large value velocity defect at mesoscale 2 is direct evidence of localised deformation.

Similar behaviour of the velocity defect can be seen in other materials. In Fig. 10.6, the dependencies of the velocity defect, the velocity variance and the spall strength on the impact velocity are shown for 38CrNi3Mo steel. At an impact velocity of ~ 308 m/s, the rate of change of the velocity defect increases.

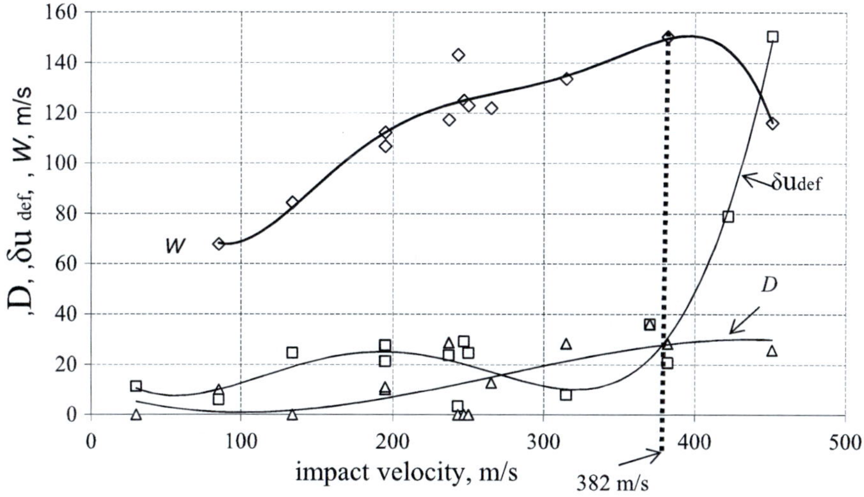

Fig. 10.6 Dependence of the velocity defect, δu_{def}, the velocity variance, D, and the pull-back velocity, W, on the impact velocity for the 38CrNi3Mo steel

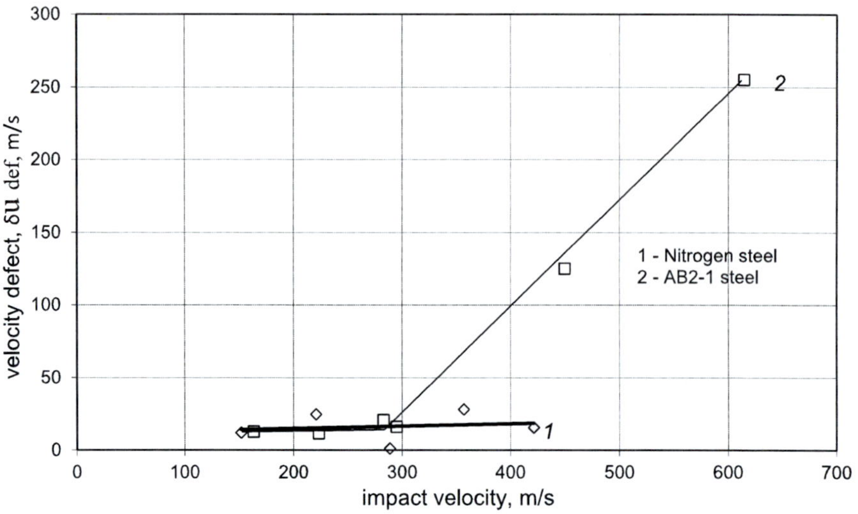

Fig. 10.7 Dependence of the velocity defect, δu_{def}, the velocity variance, D_{ms1}, and the pull-back velocity, W, on the impact velocity for nitrogen steel (1) and AB2-1 steel (2)

The transition into a structurally unstable state is sensitive to and dependent on the type of material. For comparison, in Fig. 10.7, the dependencies of the velocity defect δu_{def} on the impact velocity are provided for nitrogen steel (1) and complex alloyed AB2-1 steel (2). In AB2-1 steel, the structural transition happens at an impact velocity of 283 m/s. At the same velocity, the nitrogen steel does not undergo a structural transition: the velocity defect remains small within the overall impact velocity region.

The structural transition from mesoscale 1 to mesoscale 2 depends on the velocity distribution in mesoscale 1. As the velocity distribution is different for different mesoparticle 2 elements, the intensity of the meso–macro energy exchange and the value of the velocity defect for different points of the specimen can also differ. The effect of non-uniform dynamic straining on the initiation of the structural transitions is demonstrated in Fig. 10.8, where the free surface velocity profiles are provided for AMg-6 aluminium alloy targets loaded at impact velocities of 178 and 187 m/s. The strain rates for both tests are approximately identical. In Fig. 10.8a, the amplitude of the fringe signal gradually decreases to the plateau of the compressive pulse, which indicates the gradual widening of the velocity variance at mesoscale 1, D_{ms1}. At the same time, the number of fringe signal beats that determine the peak value of the free surface velocity, $U_{fs\ max}$, according to Eq. (9.1), corresponds to the principle of doubling the particle velocity at the free surface of the target ($U_{imp} = U_{fs\ max}$). At this point on the target, the transition to a structurally unstable state does not occur.

As for Fig. 10.8b, the free surface velocity turns out to be 71 m/s smaller than the impact velocity, that is $U_{imp} - U_{fs\ max} = 71$ m/s. The transition into a structurally unstable state happens at an impact velocity of 70 m/s (point A). These experiments

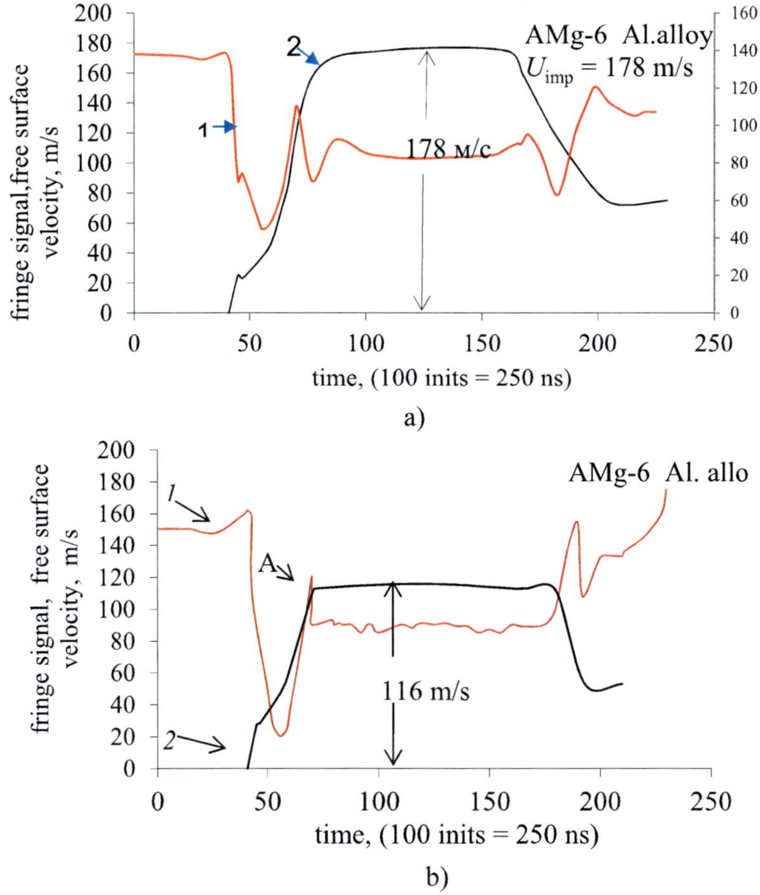

Fig. 10.8 Interference signals and time-resolved free surface velocity profiles, $u_{fs}(t)$, for the AMg-6 aluminium alloy targets loaded at an impact velocity of **a** 178 m/s and **b** 187 m/s. The break in the interference signal in the second sample is indicated by the symbol "*A*"

show that for this type of material, the transition into a structurally unstable state is a local event.

In Fig. 10.9, the dependence of the free surface velocity on the impact velocity is presented for 38CrNi3MoV steel. The dependence experiences a break at an impact velocity of 376 m/s. This impact velocity corresponds to the structural instability threshold after which the velocity defect increases. Two targets were used for the microstructural investigation of this steel: a target loaded at an impact velocity of 320 m/s and a target loaded at an impact velocity of 451 m/s. The results are presented in Fig. 10.10. In the first case, the spall zone of the target consists of the cracks parallel to the free surface of target. In this case, the fracture of the material flows in the form of cleavage, which corresponds to the uniform regime of dynamic straining

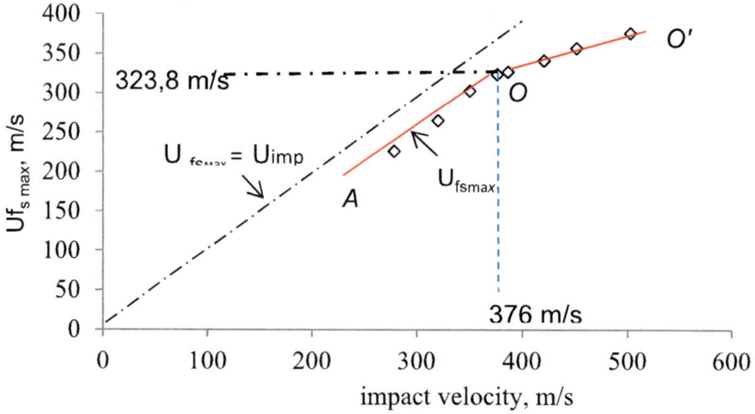

Fig. 10.9 Dependence of the maximum free surface velocity, $U_{fs\ max}$, on the impact velocity for 38CrNi3MoV steel. The dash-dotted line at 45° corresponds to the equality of the free surface velocity to the impact velocity under a symmetrical collision. The transition to a structurally unstable state occurs at an impact velocity of 376 m/s

Fig. 10.10 Microstructure of the spall zone for a 38CrNi3MoV steel target loaded at an impact velocity of **a** 320 m/s (upper image) and **b** 451 m/s (lower image)

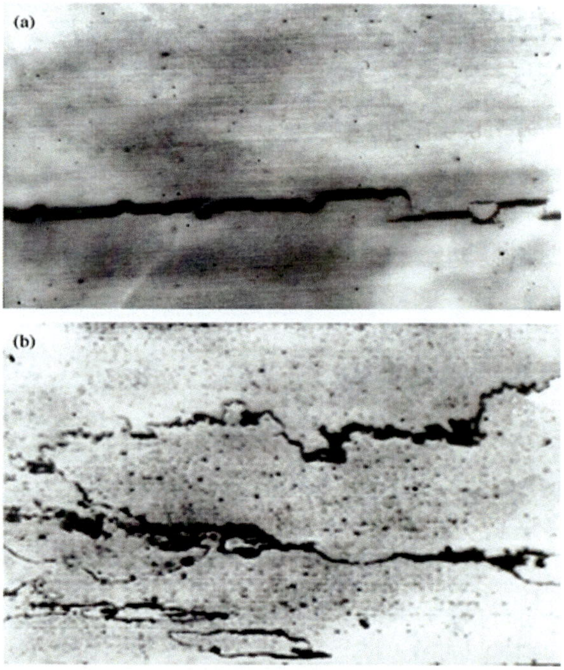

(Fig. 10.10a). In the second case, the spall zone represents the total large-scale structures restricted by closed cracks and pores (Fig. 10.10b). The formation of large-scale structures in the second case is initiated by the transition of the deformed material into a structurally unstable state.

10.5 Scenario 2: The Shock-Induced Non-equilibrium Dynamic Deformation → Quasi-Equilibrium Dynamic Deformation Transition. Large-Scale Formations at Small Spatio-Temporal Scales

Experiments on dynamic deformation show that the mechanisms of structure formation are determined not only by the initial state of the medium but also by the regimes of shock loading. An important characteristic of structure formation under shock loading is the rate of inserting the stress pulse into the material. In accordance with the modern theory of transport processes [19], under short pulse loading, the initial stage of dynamic straining is the formation of large-scale structures. At the initial stage of dynamic deformation, the response of the material to impact is elastic and the shock-induced transport of momentum and energy flows in the form of wave motions of the medium. Only after equilibrium is established the diffusion processes are initiated. Specifically, the simulation of shock wave processes in copper shows that the PVDF comes to an equilibrium state in 11.5 ns, which corresponds to a distance of ~ 50 μm [9]. A similar point of view is confirmed in the experiments of [18], where, together with the macroscopic response of material to shock, the temperature of specimens was measured using infra-red equipment. It has been shown that the common assumption that 90–95% of the kinetic energy transforms into heat in the case of dynamic deformation is not true. In this particular case, only 35–50% of the energy transforms into heat, while the rest of the energy powers the formation of large-scale structures.

Under uniform conditions of dynamic deformation, the rate of energy and momentum exchange between the macro- and meso-scales is small. Therefore, the velocity defect is constant or increases slowly (see Figs. 10.5 and 10.6). The avalanche-like increase of velocity defect signals the beginning of non-equilibrium dynamic deformation. The value of the velocity defect reflects the intensity of the meso–macro energy exchange when the valuable part of the external energy has not had time to transfer directly on the atomic scale. In this case, the shock wave energy is partly expended on structural rearrangements and the heterogenisation of the material. Under these conditions, the only possibility of observing the large-scale post-shock structures is by loading the target with short stress pulses on the order of several tens of nanoseconds.

To test the above assertions, a series of shock experiments was conducted using nanosecond durations of stress pulses. The first set of experiments with thin targets was performed using a light gas gun facility in accordance with the experimental

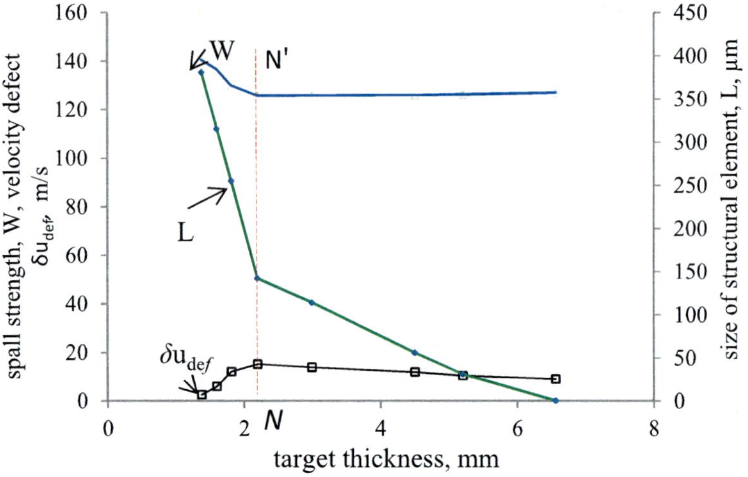

Fig. 10.11 Dependence of spall strength, W; velocity defect, δu_{def}; large-scale structure and element dimension, L, on the target thickness

technique described in Chap. 8. All the tests were carried out at the impact velocity of 405 ± 20 m/s. The thickness of targets ranged from 1.38 to 6.56 mm. The temporal free surface profile for a 2.2 mm 1565 aluminium alloy target loaded at an impact velocity of 422.8 m/s is provided in Fig. 10.4. The value of the velocity defect ($\delta u_{\text{def}} = U_{\text{imp}} - U_{\text{fs max}}$) equals ~15.3 m/s. In Fig. 10.11, together with the velocity defect, the dependence of the spall strength (or pull-back velocity), W, on the impact velocity is plotted. At target thicknesses greater than 2.2 mm, the spall strength is constant. A target thickness of 2.2 mm and a stress pulse duration of 125 ns are the critical values after which the spall strength and velocity defect do not change significantly. This shows that the intensity of the meso–macro energy exchange changes only at 2.2 mm.

The change of the velocity defect occurs in parallel with the structure formation, which affects the dynamic strength of the material, the spall strength of the material changes. Such a change in the structure formation is inherent for non-equilibrium processes. In the case of an equilibrium stage of the process, instead of forming large-scale structures, the uniform dynamic straining flows. To verify this assertion, comparative microstructural investigations of thick and thin post-shock specimens were performed. In Fig. 10.12, the microstructure patterns of the spall zones for 2.2 mm and 6.56 mm thick targets are provided to compare. In Fig. 10.12a, the closed structural formations of 100–150 μm in diameter are presented. In this case, the spall fracture happens within a contact zone of pulse loading, that is, within the region of non-equilibrium dynamic deformation. With an increased stress pulse duration, the dynamic deformation equilibrates, which leads to uniform dynamic straining of the material. In this case, the spallation occurs in the form of cleavage (see Fig. 10.12b).

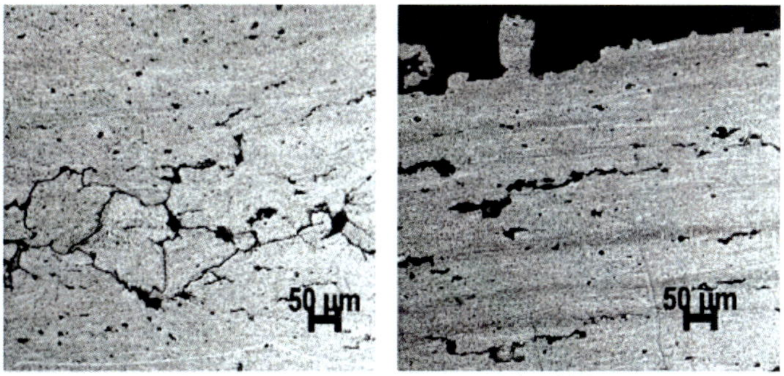

a) b)

Fig. 10.12 Spall zones in 1565 aluminium alloy: **a** the thickness of target is 2.2 mm and the stress pulse duration is 140 ns; **b** the target thickness is 6.56 mm and the stress pulse duration is 300 ns

In the second set of tests, the D16 aluminium alloy targets were loaded with stress pulses generated using a high-power electron beam. In order to avoid the influence of electron beam radiation on the material and the shock wave, the electron beam first loads the plane anode of the accelerator after which the pressure transfers into the target. The maximum value of the pressure equals approximately 4.0 GPa, while the stress pulse duration equals 40 ns. This pulse loading results in the nucleation of large-scale structures of the order of 20–50 μm. Relict structures are present due to the tension within the spall zone of the target. SEM micrographs of such structures are provided in Fig. 10.13.

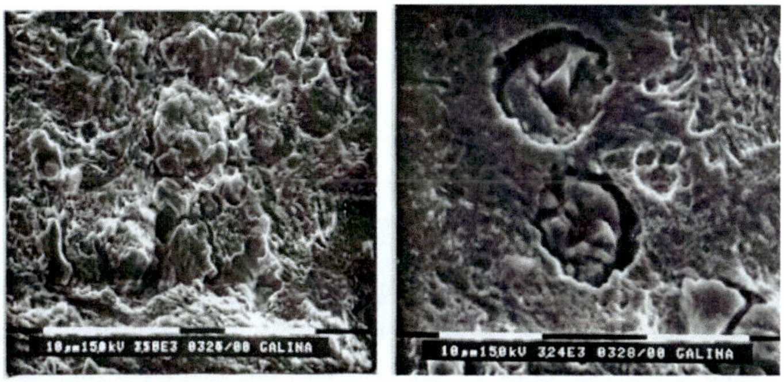

a) b)

Fig. 10.13 Relict large-scale formations in D16 aluminium alloy targets after loading with the high-powered pulse electron beam

10.6 Structural Instability Under Dynamic Compression and Resistance to High-Velocity Penetration

The instability threshold in tests under uniaxial strain conditions can be considered as independent strength characteristic of a material. This strength characteristic can be used to calculate parameters of the penetration of an elongated impactor into the target.

The penetration depth can be calculated using the Tate–Alekseevskii equation [23–26]:

$$\frac{1}{2}\rho_1\left(U_{\text{imp}} - u\right)^2 + \mathrm{Y} = \frac{1}{2}\rho_2 u^2 + R \tag{10.20}$$

Here, u is the particle velocity, U_{imp} is the velocity of impactor, and Y and R are the empirical fitting parameters determined by the dynamic strength of the impactor and the target materials. These parameters take into account the difference of the experimental penetration depth from the depth predicted by the hydrodynamic model. In the Tate–Alekseevskii model, the parameters are determined through the dynamic strength of the material, H_{d}, which is related to the dynamic yield limit Y_{d} as follows:

$$H_{\text{d}} = (3 \div 3.5)Y_{\text{d}}. \tag{10.21}$$

The latter, in turn, is determined by the Hugoniot elastic limit, σ_{hel}:

$$Y_{\text{d}} = \frac{(1-2\nu)}{(1-\nu)}\sigma_{\text{hel}}, \tag{10.22}$$

where ν is the Poison coefficient. The resistance to penetration is determined by the fracture characteristics of the material. The analysis of the shock wave processes under penetration shows that in the so-called stagnation point (the critical point of plastic flow inside the target immediately before the impactor), the one-dimensional deformation is initiated. Unlike the high-velocity penetration tests, the tests under uniaxial strain conditions are sufficiently informative. For example, the threshold for structural instability, spall strength, velocity defect and dynamic yield limit can be determined in real time. One of these characteristics, the threshold for structural instability, σ_{inst}, can be used as the target strength characteristic R in Eq. (10.20):

$$\sigma_{\text{inst}} = \frac{1}{2}\rho_2 C_{\text{p}} U_{\text{inst}}, \tag{10.23}$$

where C_{p} is the plastic front velocity measured under uniaxial strain conditions and U_{ins} is the particle velocity corresponding to the instability threshold under compression. The results of calculation by using Eqs. (10.20)–(10.23) for a set of construction materials are provided in Table 10.1. The values of R determined according to Tate's

Table 10.1 Comparison of strength parameters of resistance to high-velocity penetration into target R and the threshold for structural instability,σ_{inst}, for three types of material

Target material	σ_{inst} (GPa)	R (Tate) (GPa)	C_{p} (mm/µs)	U_{inst} (m/s)	σ (GPa)
D16 Al alloy	40	0.6	5.35	80	0.58
M2 copper	5	2.44	4.38	132	2.54
38CrNI MO V steel	75	4.01	5.00	210	3.96

method and the threshold for structural instability σ_{inst} determined in uniaxial strain experiments are seen to agree. This means that the characteristic of resistance to high-velocity penetration has a concrete physical meaning: the value of σ_{inst} determines the threshold for structural instability in the tests under uniaxial strain conditions.

10.7 The Effect of Velocity Non-uniformity on Penetration Depth

In order to account for the non-uniformity of dynamic straining on the penetration process, the particle velocity at the mesoscale can be presented in the form [27]

$$u = \overline{u} + D, \tag{10.24}$$

where $\overline{u}$ is the mean particle velocity and D is the velocity variance. The Bernoulli equation for high-velocity penetration can then be written in the following form:

$$\frac{1}{2}\rho_1[U - (\overline{u} + D)]^2 + \sigma_1 = \frac{1}{2}\rho_2(\overline{u} + D)^2 + \sigma_2. \tag{10.25}$$

Then, the mean particle velocity equals:

$$\frac{1}{2}\rho_1[U - (\overline{u} + D)]^2 + \sigma_1 = \frac{1}{2}\rho_2(\overline{u} + D)^2 + \sigma_2, \tag{10.26}$$

where $\alpha = 1 - \lambda, \lambda = \frac{\rho_1}{\rho_2}$ and $M = \frac{2}{\rho_1}(\sigma_2 - \sigma_1)$.The mean particle velocity $\overline{u}$ and penetration depth L are directly related to the velocity variance D. In order to test that hypothesis, two sets of tests were been performed. In the tests under uniaxial strain conditions, the values of the velocity variance were measured. In the second set of tests, the targets were subjected to shape charges. The results of the tests are provided in Table 10.2.

The above results reflect the dependence of the penetration depth on the non-uniformity of the dynamic deformation process. For all the materials, the greater the velocity variance, the smaller the penetration depth of the cumulative strain.

Table 10.2 Velocity variance, D, and penetration depth, L, for different steels

Kind of steel	D (m/s)	M	L (mm)
30CrNi4 Mo $\sigma_b = 800$ MPa	166	0.15	3.45
30CrNi4 Mo $\sigma_b = 960$ MPa	160	0.19	2.9
30CrNi4 Mo $\sigma_b = 1340$ MPa	136	0.205	2.2
30CrNi4 Mo $\sigma_b = 1600$ MPa	64	0.19	2.9

References

1. Duvall GE (1978) Irish J Phys Tech 7:57
2. Horsthemke W, Lefever R (1984) Noise-induced transitions. Springer, New York
3. Meshcheryakov YuI, Divakov AK, Zhigacheva NI, Makarevich IP, Barakhtin BK (2008) Dynamic structures in shock-loaded copper. Phys Rev B 78:64301–64316
4. Indeitzev DA, Meshcheryakov YuI, Kuchmin AYu, Vavilov DA (2015) Multiscale model of steady wave shock in medium with relaxation. Acta Mech 22(3):917–930
5. Hintze JO (1962) Turbulence. Graw Hill Inc., New York 560p
6. Prokuratova EI, Indeitzev DI (1995) The conditions for the dislocation distribution localization under dynamic loading. Dymat J 6(¾):229–233
7. Meshcheryakov YI, Prokuratova EI (1995) Kinetic theory of continuously distributed dislocations. Int J Solids Struct 3(12):1711–1726
8. Meshcheryakov YI (2019) Kinetic theory of dislocations and mesoscale formation in dynamically deformed solids. In: Generalized continua and structures. Advanced structured materials. Springer Nature. https://doi.org/10.1007/978-3-030-11665-1_22
9. Yano K, Horie Y (1999) Discrete-element modeling of shock compression of polycrystalline copper. Phys Rev B 59:13672–13680
10. Grady DE, Asay JR (1982) Calculation of thermal trapping in shock deformation of aluminum. J Appl Phys 83(1):7350–7354
11. Grady DE, Kipp ME (1987) The growth of unstable thermoplastic shear with application to steady-wave shock compression in solid 35(1):95–119
12. Kanel GI (2018) Shock waves in solid state physics. CPC. Taylor&Francis Group, p 215
13. Barker LM, Hollenbach RE (1974) Shock wave study of the phase transition in iron. J Appl Phys 45(11):4872–4887
14. Meshcheryakov YI, Zhigacheva NI, Divakov AK, Konovalov GV, Osokin EP (2020) Multiscale mesostructure formation under impact loading. Phys Mesomech 23:231–240. https://doi.org/10.1134/S1029959920030078
15. Duvall G.E. Propagation of plane shock waves in a stress-relaxing medium. In: *"Stress Waves in Inelastic Solids"* Eds. H. Kolsky and W. Prager. Springer-Verlag, Berlin. P. 20–32. (1964).
16. Taylor JW (1965) Dislocation dynamics and dynamic yielding. J Appl Phys 36(10):3146–3155
17. Asay JR, Chhabildas C (1981) Determination of the shear strength of shock compressed 6061-T6 aluminum. In: Meyers MA, Murr LE (eds) Shock waves and high-strain-rate phenomena in metals: concepts and applications. Plenum Publishing Corporation, New York
18. Ravichandran G, Rozakis AJ, Hodovany J, Rozakis P (2001) On the convention of plastic work into heat under shock loading. In: Furnish MD, Thadhani NN, Horie Y-Y (eds) Shock compression of condensed matter-2001. AIP Conference
19. Khantuleva TA, Meshcheryakov YI (2015) On the instability of plastic flow at the mesoscale under high-velocity deformation of solid. Fizicheskaya Mesomekhanika. 19:5–13

20. Case S, Horie Y (2007) Discrete element simulation of shock wave propagation in polycrystalline copper. J Mech Phys Solids 55:389–514
21. Belikova AF, Buravova SN, Petrov EF (2013) Localization of under dynamic loads. J Tech Phys 83(8):68. (Russian Journal)
22. Barker LM, Hollenbach RE (1974) Shock wave study of the phase transition in iron. J Appl Phys 45(11):4872–4877
23. Stilp AJ, Hohler V(1990) Long-rod penetration mechanics. High-velocity impact dynamics. Willey, New York, pp 321–404
24. Tate A (1967) Theory for deceleration of long rod penetration. J Mech Phys Solids 141
25. Tate A (1986) Long rod penetration. Part 1. Int J Mater Sci 28:535
26. Tate A (1986) Long rod penetration. Part 2. Extension of the hydrodynamic theory of penetration. Int J Mater Sci 28:599
27. Savenkov GG, Meshcheryakov YI, Barakhtin BK (2010) Influence of the velocity of non-uniformity on penetration of cumulative charge and elongated impactors. J Tech Phys 80(10):136–141. (Russian edition)

Chapter 11
Mesoscopic Criteria for the Dynamic Strength of Materials

Abstract As the macroscopic strength characteristic, the spall strength is used. This characteristic is inferred from the rise time velocity profile, which is measured in shock tests under uniaxial strain conditions. As a quantitative characteristic of structural kinetics,the velocity variation at mesoscale 1 and mesoscale 2 is applied. For all of the materials tested, including steel, copper and aluminium alloys, there is exchange of energy and momentum between different spatial scale levels.

11.1 Introduction

In this chapter, the interscale interaction is studied with two techniques: an experimental investigation of the response of different steels to shock loading and a microstructural investigation of the post-shock samples. In addition to the free surface velocity profile, which allows the spall strength of material to be determined, two additional dynamic characteristics—the particle velocity variance and defect of particle velocity—are observed. The first characteristic is the quantitative measure of the mesoparticle velocity scattering whereas the second value characterises the intensity of the interscale energy and momentum exchange. These characteristics reflect the material behaviour in real time, whereas the microstructural investigations characterise the post-shock material. Comparing the two investigations facilitates building a bridge between the physics and the mechanics of dynamic plasticity and fracture.

11.2 Mesoscale Criteria for Dynamic Strength

The experimental investigations of the coupling between the macroscopic response of a material to shock loading and the structural kinetics at mesoscale 1 and mesoscale 2 are studied with five kinds of ductile steel of different chemical compositions and different thermo-mechanical treatments: 40CrNiMo and 38CrNi3MoV armour steels, 4340 steel, 28CrNi3MoV steel and 16Cr11Ni2V2Mo steel.

Y. Meshcheryakov, *Multiscale Mechanics of Shock Wave Processes*, Shock Wave and High Pressure Phenomena, https://doi.org/10.1007/978-981-16-4530-3_11

11.2.1 40CrNiMo Steel

This is a complex alloyed high-strength steel (*HRc* 54) which is commonly used as armour steel. The spall strength for the steel was studied for two sets of targets of different thicknesses, 2 and 5 mm. These were chosen in order to check the influence of the steadiness of the wave process and the kinetics of mesostructure on spall strength. The results of the test on the 5 mm target are provided in Table 11.1.

In Table 11.1, the first column is the thickness of target, the second column is the thickness of impactor, the third column is the impact velocity, and the fourth column is the velocity defect, δu_{def}. The fifth column is the velocity variation, D_{ms1}, at mesoscale 1, and the sixth column is the spall strength (in the form of pull-back velocity), W. The dependencies of the pull-back velocity, $W = f(U_{imp})$, the velocity defect, $\delta u_{def} = f(U_{imp})$, and the mesoscale velocity variance at mesoscale 1, $D_{ms1} = f(U_{imp})$, on the impact velocity for the first set of 40CrNiMo steel targets are provided in Fig. 11.1.

Let us analyse this result from the point of view of mesostructure kinetics. In accordance with the approach we have developed, the velocity variance at mesoscale 1, D_{ms1}, characterises the intensity of the short relaxation processes in deformed material, whereas the mesoscale 2 velocity variance, D_{ms2}, characterises the scattering of the intensity of the meso–macromomentum exchange at different points on the target. The velocity scattering at mesoscale 2 characterises the fragmentation and fracture of the material. If the effect of velocity variance at mesoscale 1 is higher than that at mesoscale 2, the local stresses have time to relax before they reach the critical value for the initiation of the deformation processes at mesoscale 2. Thus, the first criterion for dynamic strength follows from the condition:

$$D_{ms1} \geq D_{ms2}. \tag{11.1}$$

Table 11.1 Results of shock tests on the 5 mm 40CrNiMo steel

H_t (mm)	h_{imp}(mm)	U_{imp}(m/s	δu_{def} (m/s)	D (m/s)	W (m/s)
5.35	1.98	211	16	4	166.1
5.30	2.00	215	14	2	168.7
5.36	1.96	290	22	7.6	199.1
5.06	2.00	307	22.5	8.5	192.3
5.59	1.95	309	23	9.3	195.8
5.25	1.98	329	24	10	195.1
5.26	2.15	35	25	11.7	197.6
5.20	1.99	379	28	16.1	200.9
5.35	1.99	405	29.5	21.5	214.1
4.83	1.97	437	30.2	19.5	200.3

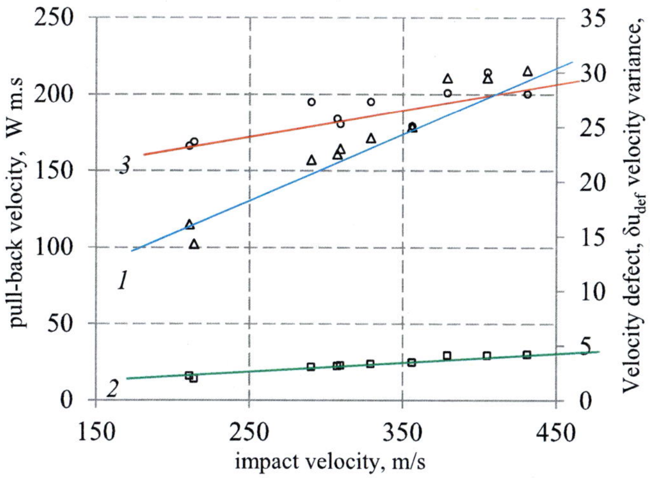

Fig. 11.1 Dependencies of the velocity defect δu_{def} (1), the velocity variance, D_{ms1}, at mesoscale 1 (2) and the pull-back velocity, W (3), on the impact velocity for the first set of 40CrNiMo steel samples

The interference technique allows the velocity variance to be measured only at mesoscale 1. In order to determine the velocity variance at mesoscale 2, one has to perform a set of shock tests for identical targets loaded within the impact velocity range of 200–450 m/s (Fig. 11.2).

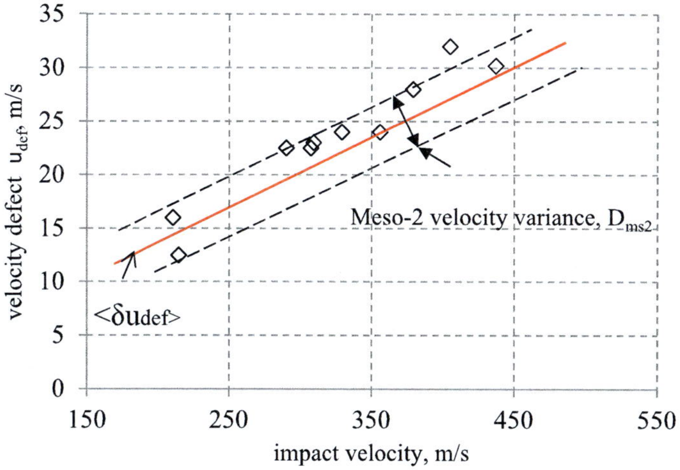

Fig. 11.2 Dependence of the velocity defect on the impact velocity for the first set of 40CrNiMo steel targets. The scattering of the velocity defect determines the value of the velocity variance at mesoscale 2, D_{ms2}

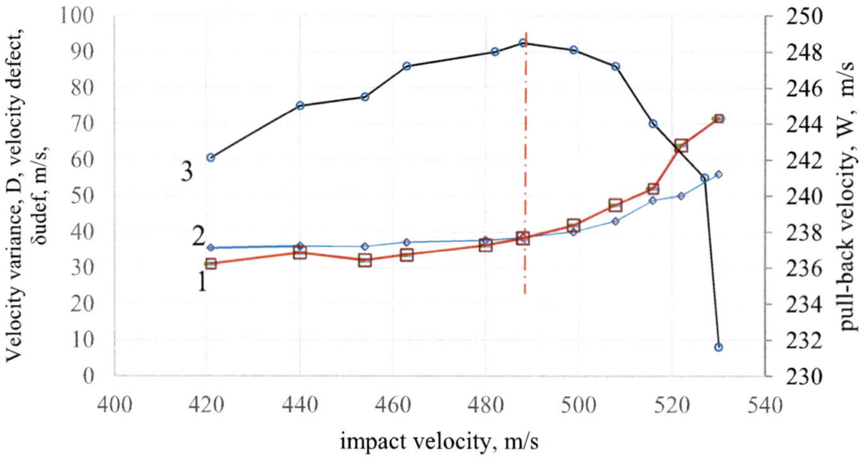

Fig. 11.3 Dependencies of the velocity variance at mesoscale 1, D_{ms1} (1), velocity defect, δu_{def} (2), pull-back velocity, W (3), on the impact velocity for the second set of 40CrNiMo steel targets

The second criterion for the optimisation of the dynamic strength of the material follows from the condition:

$$D_{\mathrm{ms1}} \geq \delta u_{\mathrm{def}}. \tag{11.2}$$

The criterion (11.2) supposes that the velocity variance at mesoscale 1 must be greater than the velocity defect.

In Fig. 11.3, the dependencies of $W = f(U_{\mathrm{imp}})$, $D_1 = f(U_{\mathrm{imp}})$, and $\delta u_{\mathrm{def}} = f(U_{\mathrm{imp}})$ for the second set of 40CrNiMo steel targets are presented together.

In Fig. 11.4, the dependence of the velocity defect on the impact velocity is plotted together with the deviations from the mean value. In Fig. 11.5, the dependencies $D_{\mathrm{ms1}} = f(U_{\mathrm{imp}})$ and $D_{\mathrm{ms2}} = f(U_{\mathrm{imp}})$ are plotted together. At the impact velocity of 265 m/s, the first curve lies higher than the second curve, which means that the criterion (11.1) is true for the second set of 40CrNiMo steel samples. As for the criterion (11.2), according to Fig. 11.3, it is satisfied at impact velocities lower than 499 m/s, above which the velocity defect becomes greater than the velocity variance. This means that the optimum conditions for the second set of the steel samples are true below an impact velocity of 499 m/s, and the spall strength decreases at higher impact velocities. Note that the maximum value of the spall strength turns out to be 248 m/s, approximately equal to the value which is expected for the first set of steel targets, 250 m/s.

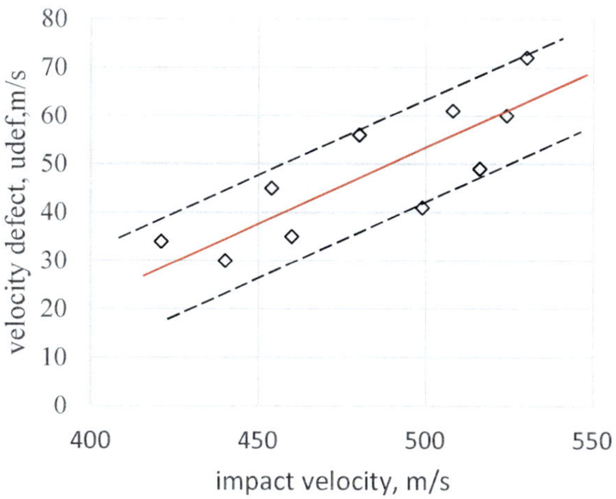

Fig. 11.4 Dependence of the velocity defect on the impact velocity for the second set of 40CrNiMo steel targets

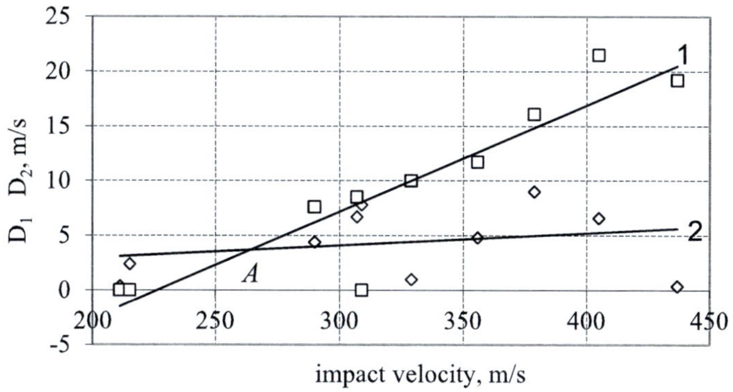

Fig. 11.5 Dependencies of velocity variance at the mesoscale 1 (1) and mesoscale 2 (2) on the impact velocity for the second set of 40CrNiMo steel

11.3 Microstructural Investigations

The standard regimen of thermo-treatment for these steels assumes low tempering, at which the structure presents a mixture of martensite and a small amount of beynite. The micrographs of spall fractures at low impact velocities show the material has failed by cleavage. Within the body of the specimen, there is a network of plane cracks of 1.5–10 μm. This spall cracking consists of parallel transverse cracks linked with short longitudinal cracks. The character of the fracture remains similar over the entire range of impact velocities studied, although at high impact velocities, the density and length of the longitudinal cracks are markedly increased.

The failure of the second set of 40CrNiMo steel samples is different. The kinematic mechanism of spall fracture is different at different impact velocities. At the lower

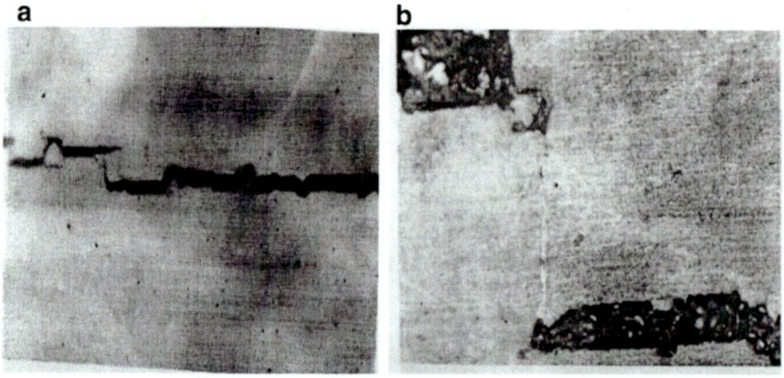

Fig. 11.6 **a** Spall crack structure for the first set of 40CrNiMo steel samples at an impact velocity of 211 m/s (magnification 32×); **b** the spall crack structure for the first set of 40CrNiMo steel samples at an impact velocity of 343 m/s (magnification 125×)

boundary of the impact velocities, the spall cracks are step-like structures with longitudinal and transverse pieces of mesoscale 2 (50–150 μm). At an impact velocity of 499 m/s, where the spall strength is at a maximum, there is a dense network of shear bands within the spall zone (Fig. 11.7a). Lastly, at an impact velocity of 521 m/s, the failure flows by means of cleavage. Such failure corresponds to behaviour of kinetic characteristics of mesostructure for that steel. At an impact velocity of 421 m/s, the values of velocity variance at mesoscale 1 and mesoscale 2 are very close to each other. In this case, the nucleation of cracks is controlled by the velocity variance at mesoscale 1. Therefore, the dimensions of the cracks turn out to be much smaller than under the condition of $D_{\rm ms1} = D_{\rm ms2}$, when the dimension of the cracks equals $L = 2C_l t_{\rm sp}$. Here, C_l is the longitudinal sound velocity and t_{sp} is the duration of spallation. For 40CrNiMo steel, $C_l = 4.71 \cdot 10^5$ cm/s and $t_{\rm sp} = 100$ ns, so the length of a single crack equals $L \simeq 500$ μm (Fig. 11.6).

At an impact velocity of 499 m/s for the second set of 40CrNiMo steel samples, $D_{\rm ms1} \simeq D_{\rm ms2}$. In the case of ductile steel, the maximum spall strength is realised by a rotational kinematic spallation mechanism. For brittle steel of high hardness, such as 40CrNiMo steel, instead of the rotational mechanism of dynamic deformation and fracture, a dense network of short adiabatic shear bands is initiated, which provides the maximum spall strength of this material.

11.3.1 38CrNi3MoV Steel

This is the complex alloyed construction steel of high strength and plastic characteristics. The presence of Ni leads to increased viscosity, whereas the presence of Mo provides high thermo-stability. In the present investigation, two sets of samples of the steel were tested. The first set of targets was made from the material as purchased,

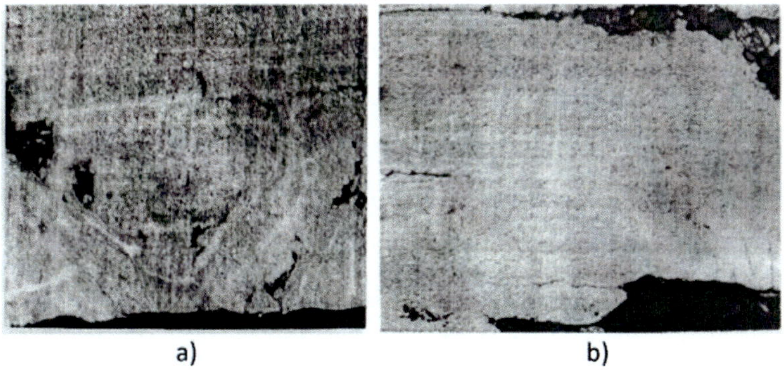

Fig. 11.7 **a** Adiabatic shear bands for the second set of 40CrNiMo steel samples at an impact velocity of 499 m/s (magnification 200 ×); **b** cleavage fracture in spall zone for the second set of 40CrNiMo steel samples at an impact velocity of 521 m/s (magnification 200×)

whereas the second set was subjected to thermo-treatment to provide homogenisation of steel structure before measurements were made.

11.3.1.1 The First Set of 38CrNi3MoV Steel Targets

In Fig. 11.8, the dependence of the velocity defect on the impact velocity together with deviations from the mean value is plotted. This allows the determination of the velocity variance at mesoscale 2. The velocity variance dependencies for mesoscale 1 and mesoscale 2 are presented in Fig. 11.9.

The velocity variance curve lies higher for mesoscale 1 than for mesoscale 2. This means that this type of steel is optimised to meet criterion (11.1). In Fig. 11.10, the

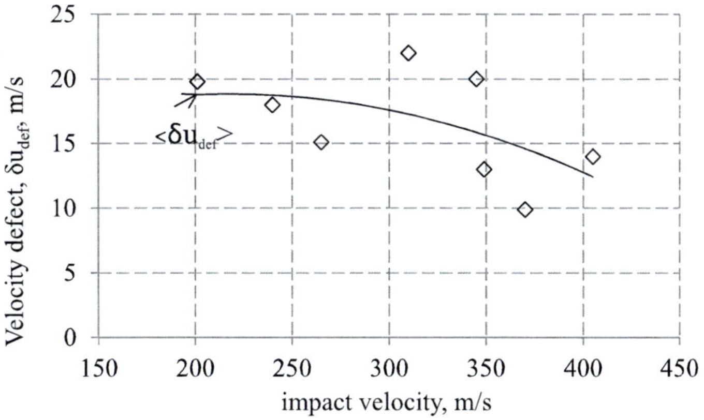

Fig. 11.8 Dependence of the velocity defect on the impact velocity for the first set of 38CrNi3MoV steel targets.

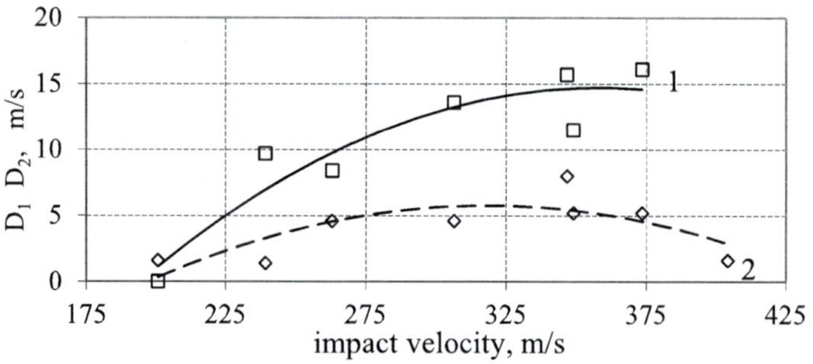

Fig. 11.9 Dependencies of the velocity variance on the impact velocity at mesoscale 1 (1) and mesoscale 2 (2) for the first set of 38CrNi3MoV steel targets

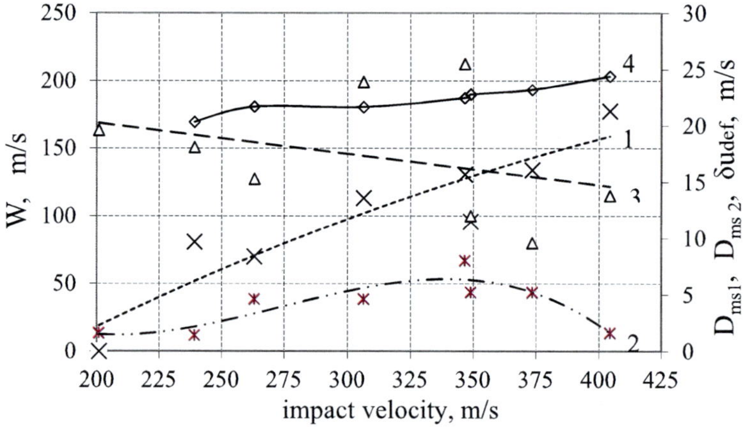

Fig. 11.10 Velocity variance at the mesoscale 1, D_{ms1} (1), mesoscale 2, D_{ms2} (2), velocity defect, δu_{def} (3) and pull-back velocity, W (4), versus impact velocity for the first set of 38CrNi3MoV steel

dependencies of velocity variance at mesoscale 1 and mesoscale 2 are plotted together with velocity defect and pull-back velocity. The curve of the spall dependence lies almost parallel to the mesoscale 1 velocity variance curve up to an impact velocity of 360 m/s (criterion (11.1)). At that impact velocity, $D_1 = f(U_{imp})$ and $\delta U_{def} = f(U_{imp})$ cross each other, so the velocity variance at mesoscale 1 becomes higher than the velocity defect (criterion (11.2)). This means that for impact velocities higher than 360 m/s, the first set of 38CrNi3MoV steel targets meets both mesoscale kinetic criteria.

11.3.1.2 The Second Set of 38CrNi3MoV Steel Targets

The results of the tests for the second set of 38CrNi3MoV steel targets are presented in Figs. 11.11, 11.12 and 11.13. This type of steel reveals non-monotonous behaviour of the spall strength. The dependence $W = f(U_{imp})$ has two regions of bending: the first region corresponds to an impact velocity of ~200 m/s and the second region to ~290 m/s (indicated with symbols $^{*)}$.

According to microstructural investigations of 38CrNi3Mo steel, the first crack within the spall zone appears at an impact velocity of 181.3 m/s. At an impact velocity of 221.2 m/s, the spall crack is solely formatted. The kinetic criterion (11.1) for this steel is accomplished within the overall range of impact velocities under investigation since the mesoscale 2 is not activated. As for criterion (11.2), at an impact velocity of 290 m/s, the velocity variance at mesoscale 2 remains higher than that at mesoscale 1. The following relationships are true:

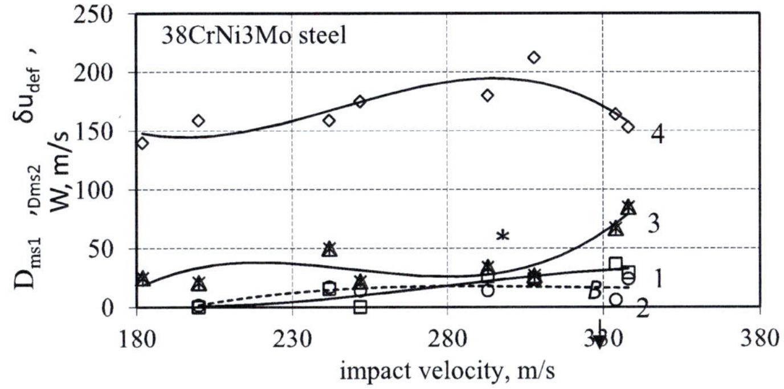

Fig. 11.11 Velocity variance at mesoscale 1 (1) and mesoscale 2 (2), the velocity defect (3) and spall strength (4) versus the impact velocity for the second set of 38CrNi3MoV steel targets

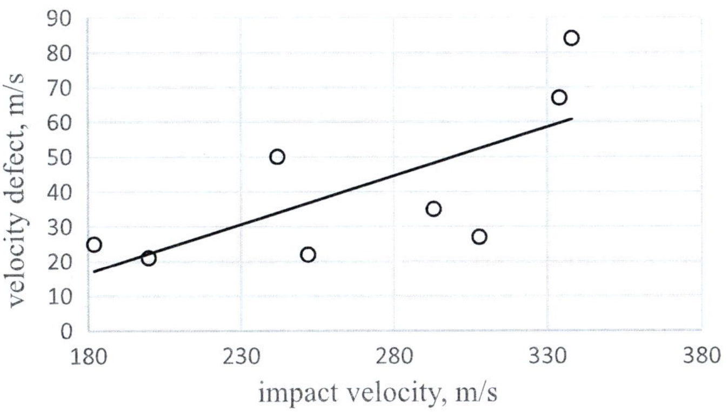

Fig. 11.12 Dependence of the velocity defect on the impact velocity for the second set of 38CrNi3MoV steel targets

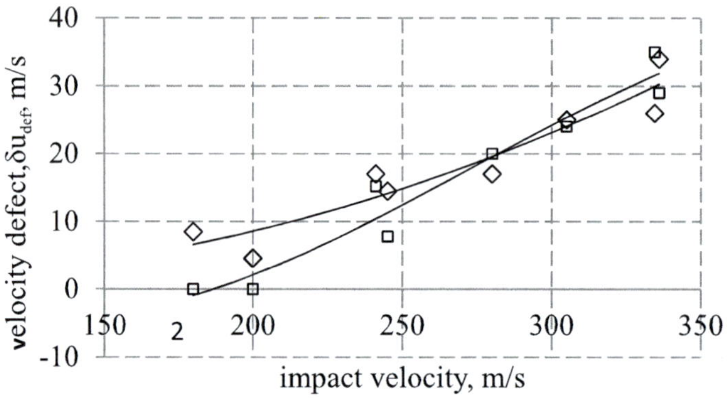

Fig. 11.13 Velocity variance at mesoscale 1 (1) and mesoscale 2 (2) versus the impact velocity for the second set of 38CrNi3MoV steel targets

$$\text{a) } D_{\text{ms1}} > D_{\text{ms2}}; \qquad \text{b) } D_{\text{ms1}} < \delta u_{\text{def}} \tag{11.3}$$

In this case, the basic mechanism of fracture is cleavage, when the horizontal dimensions of the pieces of cracked spall approximately equal the vertical dimensions. A typical picture of fracturing in that case is shown in Fig. 11.14.

As shown in Fig. 11.11, at an impact velocity of 290 m/s, the velocity variance for 38CrNi3MoV steel at mesoscale 1, D_{ms1}, becomes equal to the velocity variance at mesoscale 2, D_{ms2}. Above that impact velocity, the spall strength begins to increase. The maximum spall strength of the 38CrNi3MoV steel corresponds to an impact velocity of 290 m/s, where the dependencies $D_1 = f(U_{\text{imp}})$ and $\delta U_{\text{def}} = f(U_{\text{imp}})$ cross each other (in Fig. 11.11, this is indicated with the symbol *). Thus, the region of the second bending of the dependence $W = f(U_{\text{imp}})$ corresponds to criterion (11.2). At impact velocities over 290 m/s, the velocity defect begins to grow, while

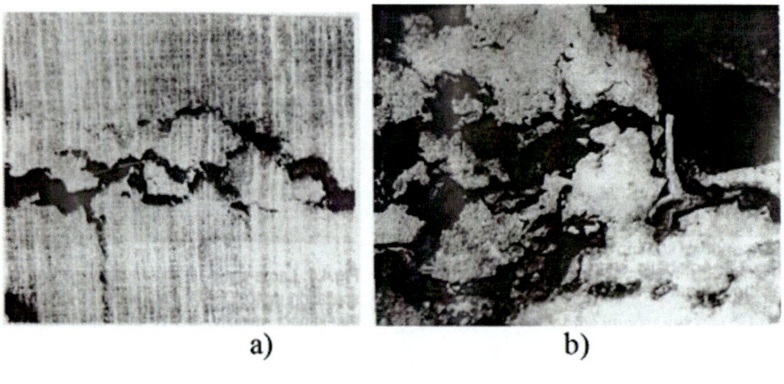

Fig. 11.14 **a** Structure of the spall cracks in a 38CrNi3MoV steel target at an impact velocity of 221 m/s (magnification 32×); **b** the evolution of spallation in a spall zone in a 38CrNi3MoV steel target after increasing the impact velocity to 390 m/s (magnification 125×)

the spall strength decreases. Thus, the spall strength is maximised at the impact velocity at which the mesoscale criterion (11.2) is true.

11.3.2 4340 Steel

This is a complex alloyed high-strength steel with the orbit structure oriented parallel to the martensite. The quasistatic strength for this steel is slightly lower compared to 40CrSNiMoV steel (*HRc* 40). The criterion (11.1) is met (see Fig. 11.16), which allows the conclusion that this kind of steel is optimised for the first kinetic criterion over the overall range of impact velocities. The spallation begins above the impact velocity of 296.4 m/s. The separate transverse cracks within the spall zone are not linked with longitudinal cracks. The latters appear only above the impact velocity of 296.4 m/s (see Fig. 11.16). In this context, the impact velocity of 296.4 m/s can be considered the threshold velocity at which the growth of longitudinal cracks begins. The length of the critical longitudinal crack equals

$$L_{cr} = D_2 \times t_{sp}, \tag{11.4}$$

where t_{sp} is the duration of spallation. This characteristic can be determined from the free surface velocity profile as the interval between the end of the plateau of the compressive pulse and the first minimum at the back front of the pulse. For the free surface profile corresponding to an impact velocity of 296 m/s, $t_{sp} = 380$ ns. The velocity variance at mesoscale 2, D_{ms2}, can be determined from the scattering of selected values of the free surface velocity presented in Fig. 11.15.

The standard deviation of the velocity for the dependence $U_{fs}^{max} = f(U_{imp})$ equals $D_{ms2} = 8.8$ m/s, and the critical length of the shear crack equals $L_{cr} = 3.44$

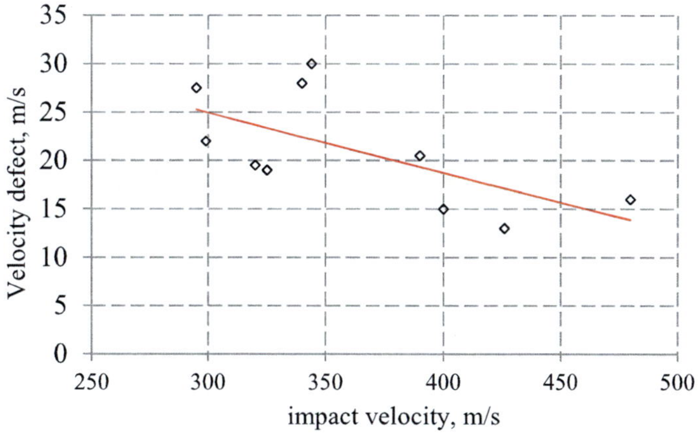

Fig. 11.15 Velocity defect versus impact velocity for 4340 steel

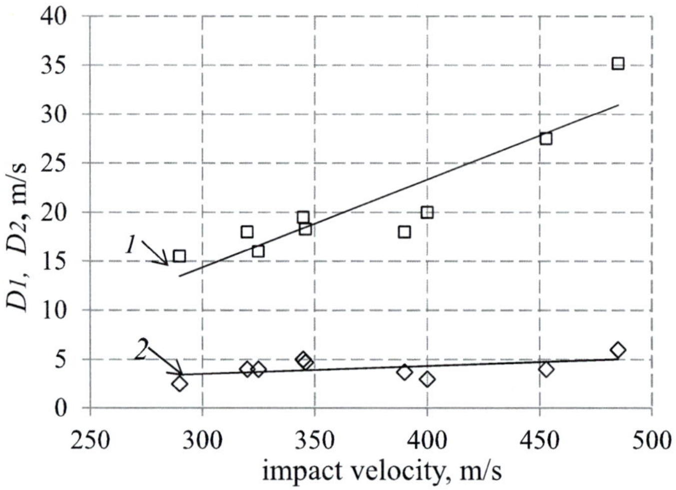

Fig. 11.16 Velocity variance at mesoscale 1, D_{ms1} (1), and mesoscale 2, D_{ms2} (2), versus impact velocity for the 4340 steel

µm. It follows from the dependencies $D_{ms1} = f(U_{imp})$, $D_{ms2} = f(U_{imp})$ and $\delta U_{def} = f(U_{imp})$ presented in Fig. 11.17 that below an impact velocity of 291 m/s, the mesoscale 1 velocity variance is greater than the mesoscale 2 velocity variance, $D_{ms1} > D_{ms2}$.

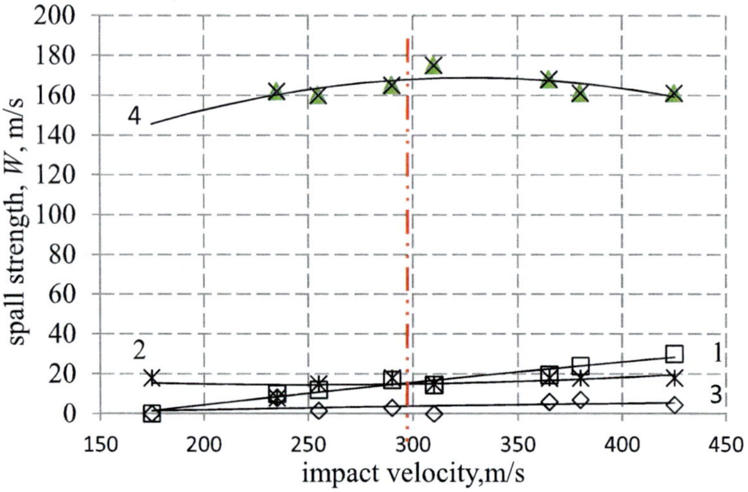

Fig. 11.17 Velocity variance at mesoscale 1, D_{ms1}, (1), and mesoscale 2, D_{ms2}, (2), velocity defect (3) and pull-back velocity, W (4), versus impact velocity for the 16Cr11Ni2V2MoV steel

11.3.3 16Cr11Ni2V2MoV Steel

This kind of steel has increased content of Cr, *W*, Mo and *B*, and is commonly used for manufacturing the parts of gas turbines. Along with high strength, this kind of steel has a high thermo-resistance. The results of our tests are illustrated in Fig. 11.17. Criterion (11.1) is seen to be true within the overall range of impact velocities studied. Accordingly, the spall strength reveals the monotonous behaviour within the overall range of impact velocities.

11.3.4 28Cr3CNiMoV Steel

The fifth type of steel is the complex alloyed 28Cr3NiMoV steel after thermal treatment (*HRc* 24). The dependencies of the pull-back velocity, $W = f(U_{\mathrm{imp}})$, the mesoscale 1 velocity variance,$D_{\mathrm{ms1}} = f(U_{\mathrm{imp}})$, and the velocity defect, $\delta u_{\mathrm{def}} = f(U_{\mathrm{imp}})$, for the 28Cr3NiMoV steel are provided in Fig. 11.18. The velocity defect is seen to lie above the mesoscale 1 velocity variance, such that $\delta u_{\mathrm{def}} > D_{\mathrm{ms1}}$, so the criterion (11.2) is not fulfilled. This means that conditions for quick stress relaxation at mesoscale 2 are not attained, which results in the nucleation of cracks and a low value of spall strength (75 m/s as compared to 250 m/s for 40CrNiMo steel). This type of steel is used in a tempered state, which provides a high plasticity (*HRc* 24). This steel is not optimised on criterion (11.2). Therefore, the mechanism of cleavage dominates within the overall range of impact velocities. A picture of typical fracturing for this steel is presented in Fig. 11.19.

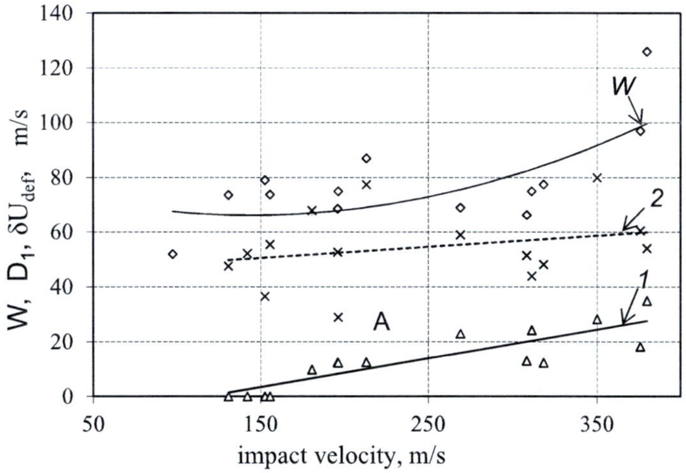

Fig. 11.18 Dependence of spall strength, W, mesoscale 1 velocity variance, D_1 (1), and velocity defect, δu_{def} (2), on the impact velocity of 28Cr3NiMoV steel

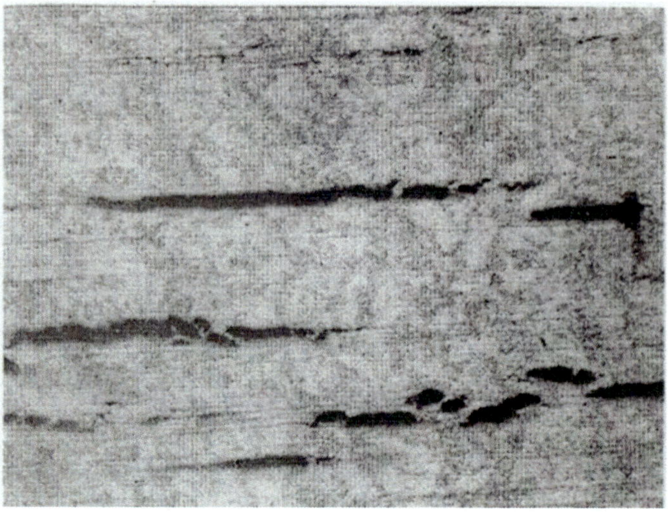

Fig. 11.19 Cleavage cracks within the spall zone of 28Cr3NiMoV steel

11.4 Analysis of the Strength Behaviour of Steels

In Fig. 11.20, the dependencies of the pull-back velocity on the impact velocity are presented for all the steel targets tested. All of the steel targets, except the second set of 38CrNi3MoV steel targets, show a monotonous dependence of spall strength on the impact velocity. For two types of steel, the first set of 38CrNi3MoV

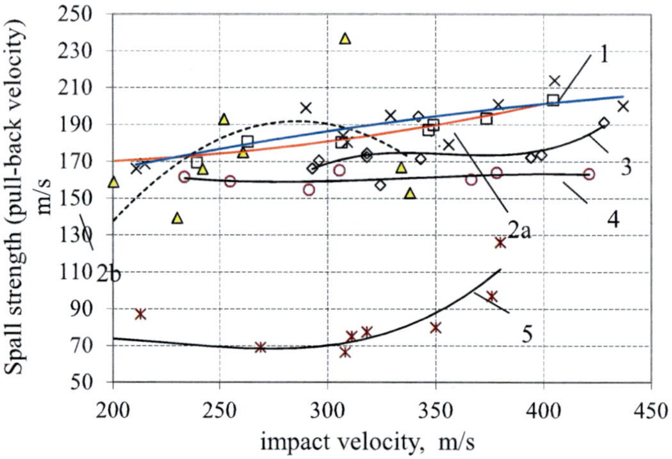

Fig. 11.20 Dependencies of spall strength on the impact velocity: 1—40CrCNiMoV; 2—38CrNi3MoV steel: (*2a*)—set 1, (*2b*)—set 2; 3—4340 steel; 4—16Cr1Ni2V2MoV steel; and 5—28Cr3NiMoV

steel targets and the first set of 40CrNiMoV steel targets, these dependencies effectively coincide, although they have different elastic precursors—60 m/s for 38CrNi3MoV steel and 120 m/s for 40CrNiMoV steel, respectively. The steels also have different quasistatic strengths—1450 MPa for 38CrNi3MoV steel and 1870 MPa for 40CrNiMoV steel, respectively. Last, according to the microstructural data, the spall fracture of 38CrNi3MoV steel flows in a viscous manner, whereas the fracture of 40CrNiMoV steel is purely brittle. The values of pull-back velocity are taken at the impact velocity at which criteria (11.1) and (11.2) are accomplished. In the case, when criteria (11.1) and (11.2) are accomplished, the dynamic strength of the material correlates with the quasistatic strength. For the 40CrNiMoV and 28CrNi3MoV steels, the expected correlation of strength characteristics is absent. The dynamic strength for these materials is determined by the behaviour of the kinetic characteristics at mesoscale 1 and mesoscale 2.

11.5 The Meso–macro-energy Exchange and Spallation

The meso–macro-energy exchange in the form of Eqs. (11.4–11.6) can be used to explain the non-monotonous behaviour of spall strength depending on the strain rate. The power balance in the spall zone can be written in the form:

$$\frac{1}{2}\rho C_0 u^2 = \frac{1}{2}\mu\frac{(u+\delta u_{\text{def}})}{h}u + \frac{1}{2}\mu\frac{D}{h}u. \tag{11.5}$$

Here, ρ is the density of the material, C_0 is the velocity of the shock wave, μ is the dynamic viscosity of the deformed material, and u is the particle velocity. The left-hand side term of Eq. (11.6) characterises the power which is brought into spall zone from the load wave. The first term in the right-hand side of the equation describes the loss of power due to the normal rupture of the material due to tension in the spall zone and reflects the motion of the spall surfaces in opposite directions. The particle velocity changes owing to the meso–macro-energy exchange mechanism, so the mean particle velocity for this term includes the velocity defect δu_{def}. The total strain rate in the spall zone due to normal tension equals:

$$\frac{d\varepsilon_1}{dt} = \frac{u+\delta u_{\text{def}}}{h}, \tag{11.6}$$

where h is the width of the spall zone. The conditions for the continuity of the deformed medium require that the particle velocity distribution at the mesoscale causes the 3D motions of the medium at that scale, such as local rotations.

The second term in the right-hand side of Eq. (11.7) characterises the power lost due to the formation of a rotation at the mesoscale at the strain rate $\frac{d\varepsilon_2}{dt} = \frac{D}{h}$. From Eq. (11.7), the spall width h can be written in the form:

$$h = \left(\frac{\rho C_0}{\mu} - \frac{\delta u_{\text{def}} + D}{h}\frac{1}{u}\right)^{-1}. \tag{11.7}$$

The time for spallation can be defined as

$$\tau_f = \frac{h}{u} = \left(\sigma - \mu\frac{\delta u_{\text{def}} + D}{h}\right)^{-1}, \tag{11.8}$$

where $\sigma = \rho C_0 u$ is the normal stress of the load wave. Equation (11.9) can be rewritten in the form:

$$\left(\sigma - \mu\frac{\delta u_{\text{def}} + D}{h}\right)\tau_f = 1. \tag{11.9}$$

The dynamic generalisation of this equation,

$$\int_0^{\tau_f} \left(\sigma - \mu\frac{\delta u_{\text{def}} + D}{h}\right)\text{d}t = 1, \tag{11.10}$$

can be considered as a dynamic fracture criterion that takes into account both the particle velocity distribution at the mesoscale and the velocity defect resulting from the meso–macro-energy exchange. Equation (11.11) determines the maximum value of the normal stress necessary for dynamic fracture in accordance with the integral criterion:

$$\int_0^{\tau_f} \sigma \text{d}t = 1. \tag{11.11}$$

In Figs. 11.21, 11.22, 11.23 and 11.24, the dependencies of the pull-back velocity, velocity variance and velocity defect are presented for different materials, including different types of steel and copper. All dependencies display their maximum spall strength at the impact velocity where the velocity variance and velocity defect equal each other (indicated by red dash-dotted lines) (Tables 11.2, 11.3 and 11.4).

11.6 Conclusions

Theoretical analysis of spallation process and numerous experiments on shock loading of different materials show that maximum spall strength of materials corresponds to equality the velocity variation and velocity defect, i.e. $D = -\delta U_{\text{def}}$.

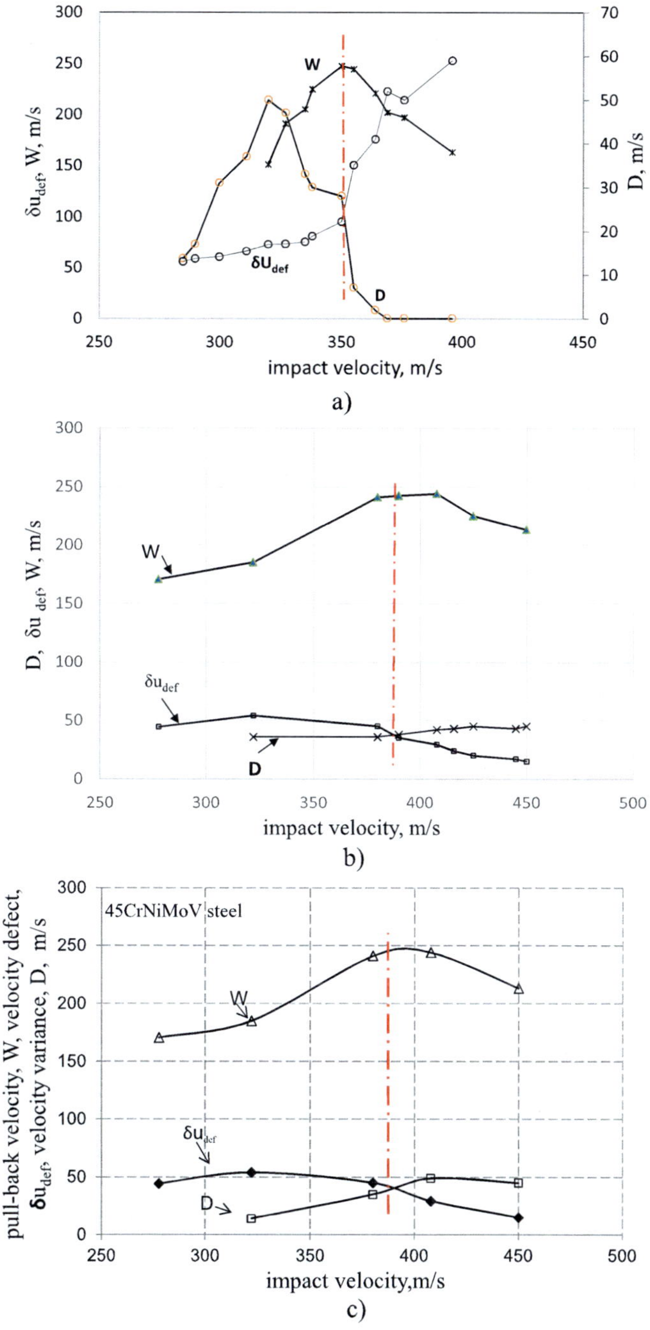

Fig. 11.21 Dependencies of the spall strength (pull-back velocity), W, the velocity defect, δu_{def}, and the velocity variance, *D*, on the impact velocities of: **a** 2Cr-2Ni-Mo-V steel; **b** Cr-2Ni-Mo-V steel; and **c** Cr-4Ni-Mo steel

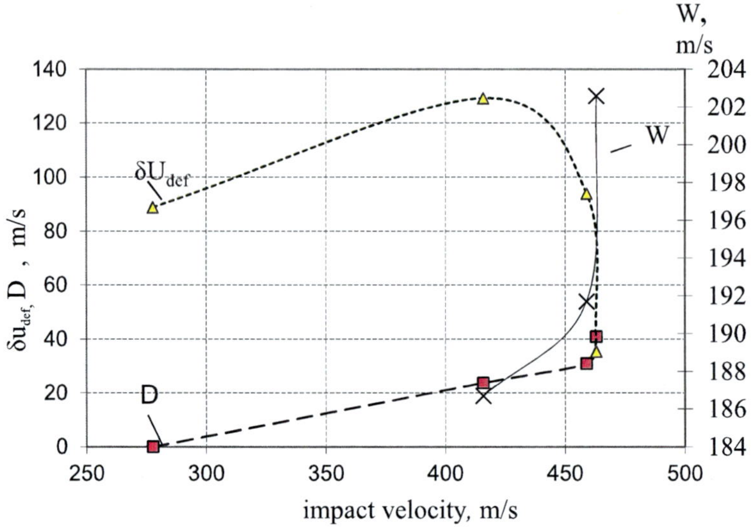

Fig. 11.22 Velocity distribution width at mesoscale 1, D; the free surface defect, δu_{def}; and the pull-back velocity, W, versus the impact velocity for 45CrNiMoV steel

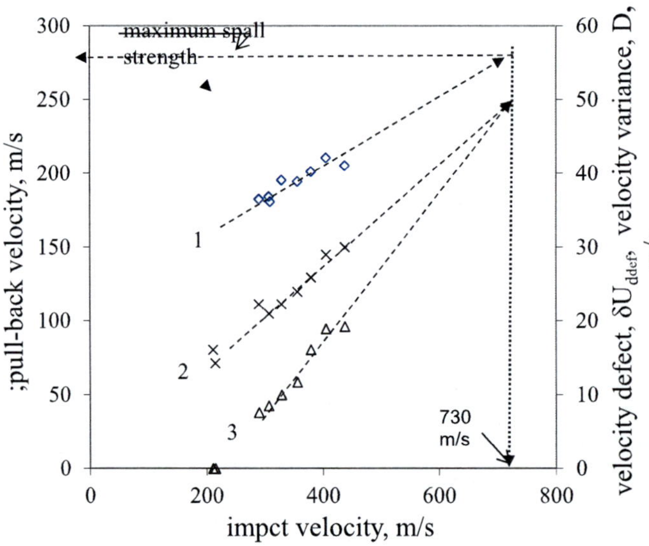

Fig. 11.23 Families of dependencies for the pull-back velocity, W (1), velocity defect, δu_{def}, (2), and velocity variance, D (3), on the impact velocity

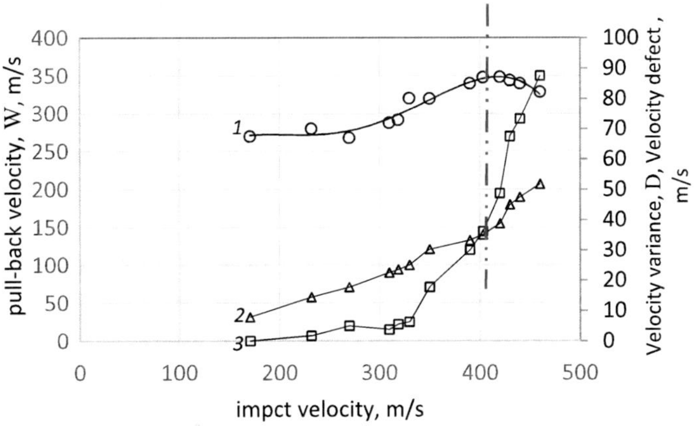

Fig. 11.24 Dependencies of spall strength, W (1), velocity defect, δu_{def} (2), and velocity variance, D (3), on the impact velocity of copper M3

Table 11.2 Chemical composition of steels

Steel grade	Chemical composition (%)							
	C	Si	Mn	Cr	Ni	Mo	V	Fe
2Cr-2Ni-Mo-V	0.3	0.25	0.45	1.7	1.7	0.8	0.12	Rest
Cr-2Ni-Mo-V	0.34	0.2	0.4	1.3	1.7	0.85	0.13	
Cr-4Ni-Mo	0.3	0.2	0.4	1.3	3.7	0.8	-	

Table 11.3 Heat treatment of steels

Steel grade	Heat treatment	Hardness
2Cr-2Ni-Mo-V	$t_{qI.} = 950$ °C, 2 h + $t_{t.} = 625$ °C, 4 h + $t_{qII.} = 825$ °C, 2 h + $t_{t.} = 590$ °C, 2 h	398 HB
Cr-2Ni-Mo-V	$t_{\text{air q.}} = 900$ °C, 3 h + $t_{qII.} = 825$ °C, 3 h + $t_{t.} = 590$ °C, 4 h	388 HB
Cr-4Ni-Mo	$t_{q.} = 900$ °C, 3 h + $t_{t.} = 550$ °C, 3 h	39 HRC

where t_q is quenching temperature, t_t is tempering temperature and $t_{\text{air,q}}$ is air quenching

Table 11.4 Pull-back velocity and rotation number

Steel grade	U_{imp}	W_{sp}	N_r
2Cr-2Ni-Mo-V	262	150	
	335	225	
	350	245	
	373	241.8	
Cr-2Ni-Mo-V	278	197.5	145
	376	235	270
	423	245	119
	450	213.1	23

Chapter 12
A Mesoscale Approach to Dynamic Recrystallisation

Abstract To confirm the effect of the velocity non-uniformity at the mesoscale on the dynamic recrystallisation process, shock tests of D16 aluminium alloy and 38CrNi3MoV steel were conducted within impact velocity strain rates of 10^5–10^7 s^{-1}. The tests were carried out under two regimes of uniaxial straining: single loading and double loading under the condition of reloading. The experiments show that dynamic recrystallisation is absent in single shock loading and present in reloading experiments. In this situation, a new scenario for shock-induced dynamic recrystallisation is suggested. The process of nucleation of the 3D structure is presented as a successive stage of rebuilding the initially chaotic dislocation structure into a polarised structure and resonance mixing.

12.1 Introduction

In this chapter, the role of the mesoscale velocity distribution is considered from the position of its influence on dynamic recrystallisation. Traditionally, dynamic recrystallisation is considered as the nucleation and growth of new grains along the initial grain boundaries of a deformed material [1]. Experimental investigations of localised deformation in copper, tantalum, titanium and brass [2, 3] have confirmed the presence of a rotational mechanism for dynamic recrystallisation. Meyers and Pak [4] have shown for titanium and Hynes for copper [5] that the nucleation time for new grains of 0.1 μm diameter, in the case of the rotational mechanism of dynamic recrystallization, is smaller by several orders of magnitude than the migration mechanism. At the initial stage of deformation, the chaotically distributed dislocations form the cell structure. With increasing deformation, the cell structure transforms into sub-grains elongated in the load direction. As the different regions of the crystal are rotated in different directions, the conditions for the geometrically necessary grains arise [6–10]. In the case of copper, the nucleation time of new grains equals approximately half a millisecond [11]. Here, atomic flows along the grain boundaries are realised with high velocities which are an order of magnitude greater compared to the flows inside the grains. For $T/T_m = 0.5$ (T_m is the melting temperature), the

Y. Meshcheryakov, *Multiscale Mechanics of Shock Wave Processes*, Shock Wave and High Pressure Phenomena, https://doi.org/10.1007/978-981-16-4530-3_12

activation energy for grain boundary diffusion equals approximately half of the activation energy of intergrain diffusion. As a result, the rotation mechanism of dynamic recrystallisation flows in 10^{-4}–10^{-5} s, which is two orders of magnitude shorter than the time for the grain boundary mechanisms.

Nevertheless, this time interval is much longer than the duration of the stress pulse initiated by shock wave loading. In this situation, alternative mechanisms to describe the shock-induced dynamic recrystallisation are needed. One such mechanism is the structural heterogeneity of the shock-deformed material. This heterogeneity can be caused by both the technology (the so-called metallurgical heterogeneity) and the formation of a heterogeneous structure because of the nonlinearity of the shock wave process. Specifically, the non-uniformity of the shock-induced velocity increases the probability of localised deformation and the local temperature, which also activates the dynamic recrystallisation process.

In the present chapter, the influence of the mesoscale velocity distribution is considered to be the driving mechanism for dynamic recrystallisation resulting from shock deformation processes.

12.2 The Reloading Regime as a Matter for Providing the Mesoscale Scenario for the Dynamic Recrystallisation

To check the role of the mesoscale non-uniformity on the shock-induced dynamic recrystallisation, shock tests were performed on D16 aluminium alloy and 38CrNi3MoV steel in single and reloading regimes. These tests revealed dynamic recrystallisation only in the second case. The reloading regime provides a tenfold increase of the duration of the velocity non-uniformity compared to the single shock case, which is sufficient to meet the conditions: $\gamma \geq 3$, $d\gamma/dt \geq 10^4\ s^{-1}$, $T \geq 0.4\ T_m$. The non-uniformity of the particle velocity is thought to be a determinant in the dynamic recrystallisation process due to the development of collective nanoshears and the discrete disorientation of grains.

To confirm the effect of the velocity non-uniformity at the mesoscale on the dynamic recrystallisation process, shock tests of D16 aluminium alloy and 38CrNi3MoV steel were conducted within impact velocity strain rates of 10^5–10^7 s^{-1}. The tests were carried out under two regimes of uniaxial straining: single loading and double loading under the condition of reloading. In the first case, the loading was performed using a 2-mm-thick impactor and 5–mm-thick targets. In the second case, the impactor consisted of two plates stuck together, the first of which was an aluminium plate 0.37–1.49 mm thick, whereas the second plate was a 15-mm-thick St 10 steel plate. Shock loadings were administered using a light gas gun of with a barrel diameter of 37 mm. The free surface velocity profiles were recorded using the velocity interferometer in accordance with the instrumentation described in Chap. 9. Structural investigations of post-shock specimens were carried out using an Axio-Observier Z-1 m optical microscope.

12.2.1 D-16 Aluminium Alloy

In Fig. 12.1, the structural states of D16 aluminium alloy specimens after single shock loading and after the reloading regime are presented. In the second case, the structure of the material is seen to have obvious features of dynamic recrystallisation with a mean grain size of ~1 μm. The conditions necessary for dynamic recrystallisation are known as follows: deformation $\varepsilon \geq 3$, strain rate $d\varepsilon/dt \geq 10^4\ c^{-1}$ and temperature $T \geq 0.4\ T_m$. Let us verify whether the above conditions are complied with in the case of single and double shock loading for D16 aluminium alloy. To determine the degree of velocity non-uniformity of the deformation process at the mesoscale, we use the free surface velocity profiles, $u_f(t)$, and the velocity variance, $D(t)$, presented in Fig. 12.2. The physical characteristics of the material and the parameters of shock loading are provided in Table 12.2. (Table 12.1).

The local strain rate in a shear band equals the ratio of the velocity mesoscale variance D, and the width of shear band is determined using $\dot{\gamma} = D/d$, where the width of the shear band is determined by the thermo-conductivity coefficient and the time Δt for the nonzero velocity variance value. The latter is determined from the time-resolved profile:

$$d = \sqrt{\chi \frac{\Delta t}{2}}.$$

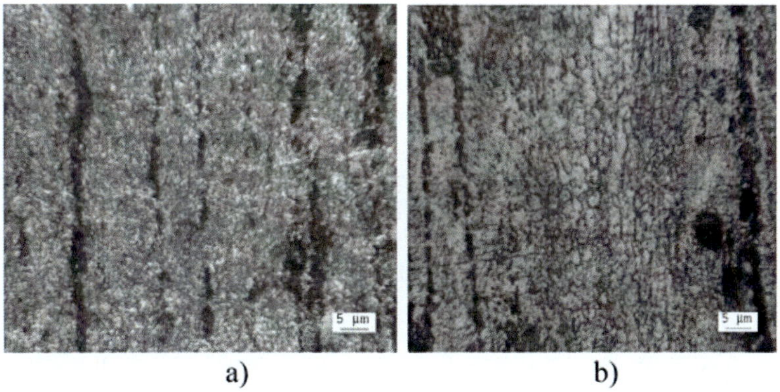

Fig. 12.1 Structural state of D16 alumunum alloy spesimens after loading in regime of single shock (a), and after loading in regime of double shock (b). The recrystalized structure in the second case is evident

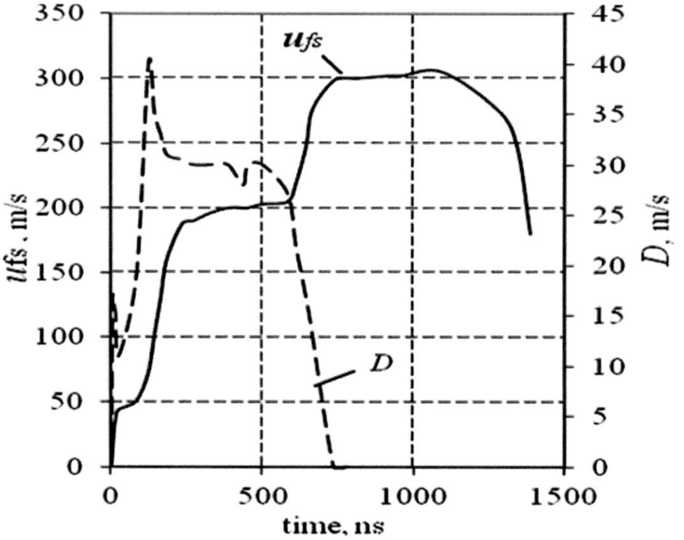

Fig. 12.2 Free surface velocity profiles, $u_{fs}(t)$, and velocity variance, D, for the double loading of a 5 mm D16 aluminium alloy target

Table 12.1 Physical characteristics and shock wave parameters for D16 aluminum alloy

Density, ρ (g/cm^3)	2.77
Specific thermos-capacity at constant volume, C_V (J/kg rad)	0.88×10^3
Velocity of longitudinal plastic wave (m/s)	6400
Velocity of transverse plastic wave (m/s)	3080
Thermo-conductivity, λ (W/m grad)	209.3
Activation energy for self-diffusion, Q (KJ/mol)	144.4

From the presented velocity profiles, this time equals $\Delta t_1 = 60$ ns for the single loading and for the double loading, $\Delta t_2 = 650$ ns; whereas, the value of the velocity variance is approximately identical for both cases: $D = 30$ m/c. The thermos-conductivity coefficient for the D16 aluminium alloy equals

$$\chi = \frac{\lambda}{C_V \rho} = \frac{209.3}{0.88 \times 10^3 \times 2770} = 85.9 \times 10^{-6} \mathrm{m}^2/\mathrm{s}$$

where λ is the thermo-conductivity coefficient, C_V is the specific thermo-capacity at a constant volume, and ρ is the density of the material.

For the single and double shock loadings, the width of the shear band equals (index 1 relates to single loading and index 2—to double loading):

$$d_1 = \sqrt{\chi \frac{\Delta t_1}{2}} = \sqrt{\frac{85.9 \times 10^{-6} \times 0.06 \times 10^{-6}}{2}} = 1.6 \times 10^{-6}\text{m};$$
$$d_2 = \sqrt{\chi \frac{\Delta t_2}{2}} = \sqrt{\frac{85.9 \times 10^{-6} \times 0.65 \times 10^{-6}}{2}} = 5.28 \times 10^{-6}\text{m}$$

The rate of shear deformation is

$$\dot{\gamma}_1 = \frac{D_1}{d_1} = \frac{29.7}{1.6 \times 10^{-6}} = 1.86 \times 10^7\text{s}^{-1}$$
$$\dot{\gamma}_2 = \frac{D_2}{d_2} = \frac{30}{5.28 \times 10^{-6}} = 5.68 \times 10^6\text{s}^{-1}.$$

The value of the shear deformation is

$$\gamma_1 = \dot{\gamma}_1 \times \Delta t_1 = 1.875 \times 10^7 \times 6 \times 10^{-8} = 1.125;$$
$$\gamma_2 = \dot{\gamma}_2 \times \Delta t_2 = 5.68 \times 10^6 \times 0.65 \times 10^{-6} = 3.7$$

And the local shear stress is

$$\tau = \rho \times C_l \times D = 2.77 \times 6.4 \times 10^5 \times 0.3 \times 10^4 = 0.532 \text{ GP}.$$

The temperature growth within the shear band is

$$\Delta T_1 = \frac{\dot{\gamma}_1 \cdot \tau \cdot \Delta t_1/\rho C_v}{1 + \chi \cdot \Delta t_1/d_1^2}$$
$$= \frac{1.78 \times 10^6 \cdot 0.532 \times 10^9 \times 0.065 \times 10^{-6}/2.77 \times 10^3 \cdot 0.88 \times 10^3}{1 + 85.9 \times 10^{-6} \cdot 0.065 \times 10^{-6}/\left(1.67 \times 10^{-6}\right)^2} = 84 \text{ K}$$

$$\Delta T_2 = \frac{\dot{\gamma}_2 \times \tau \times \Delta t_2/\rho C_v}{1 + \chi \times \Delta t_2/d_2^2}$$
$$= \frac{5.68 \times 10^6 \times 0.532 \times 10^9 \times 0.65 \times 10^{-6}/2.77 \times 10^3 \times 0.88 \times 10^3}{1 + 85.9 \times 10^{-6} \times 0.65 \times 10^{-6}/\left(5.28 \times 10^{-6}\right)^2}$$
$$= 268 \text{ K}^0.$$

The deformation conditions for dynamic recrystallisation are expressed through the temperature-compensated deformation, or Zener–Hollomon parameter [12] the rotational mechanisms play the fundamental role:

$$Z = \dot{\gamma} \exp \frac{Q}{R \times T} = 5.68 \times 10^6 \exp \frac{84 \times 10^3}{8.31434 \times 288} = 9.76 \times 10^{21},$$

where Q is the activation energy of grain boundary evolution.

The average grain size can be estimated from the formula $d_s \infty \dot{\gamma}^{-\frac{1}{2}}$. This yields A value for d of 1.17 μm, which agrees with the average grain size in Fig. 12.2b. These calculations show that positive conditions for dynamic recrystallisation are accomplished only for the shock reloading regime, which relates to the tenfold increase of the duration of the local shear processes. The intensity of these processes is determined by the duration of the mesoparticle velocity distribution process. In the case of single shock loading, the value of the local shear deformation is too small ($\gamma = 1.125$), and the temperature within the shear band equals $\Delta T = 81\ K^0$, which is certainly insufficient to initiate the dynamic recrystallisation process.

The maximum duration of the local shear processes due to the non-uniformity of the dynamic straining for all the tests did not exceed 650 ns, which is much shorter than the time necessary for the activation of the dislocation migration mechanisms for the grain boundaries. Therefore, it may be concluded that in the case of shock-induced dynamic recrystallisation, the rotational mechanisms play the fundamental role[1–4].

In order to reveal the dependence of the dynamic recrystallisation on the mesoscale velocity variance, an additional experiment at an impact velocity of 225 m/s was performed (see Table 12.2). The free surface velocity profile and velocity variance profile for one of tests is presented in Fig. 12.3.

The width of the localised zone equals

$$d_3 = \sqrt{\chi \frac{\Delta t_2}{2}} = \sqrt{\frac{85.9 \times 10^{-6} \times 0.64 \times 10^{-6}}{2}} = 5.24 \times 10^{-6}\ \text{m},$$

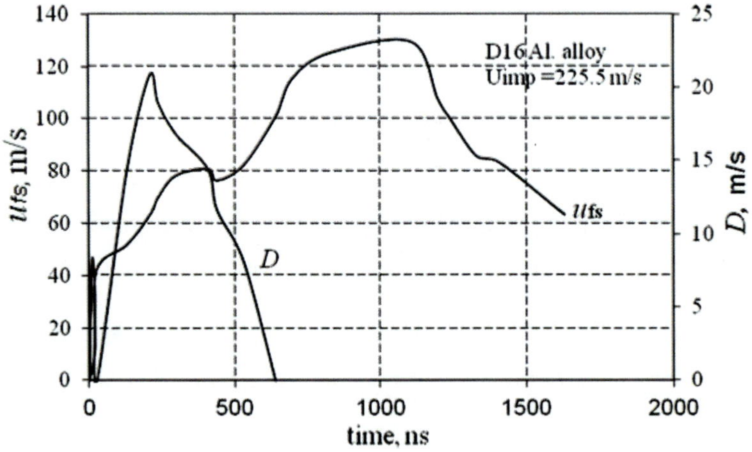

Fig. 12.3 Time-resolved free surface velocity profiles, $u_{fs}(t)$, and velocity variance profiles, D, for a D16 aluminium alloy target loaded at an impact velocity of 225.5 m/s

and the rate of shear deformation and shear stress is

$$\dot{\gamma}_3 = \frac{D_3}{d_3} = \frac{20.8}{5.24 \times 10^{-6}} = 3.97 \times 10^6 \mathrm{s}^{-1},$$

$$\gamma_3 = 3.97 \times 10^6 \times 0.64 \times 10^{-6} = 2.54,$$

$$\tau_3 = \rho C_l D_3 = 2.77 \times 6.4 \times 10^5 \times 0.208 \times 10^4 = 0.368 \text{ GPa}.$$

The temperature inside the shear band is

$$\Delta T_3 = \frac{\dot{\gamma}_3 \times \tau_3 \times \Delta t_3 / \rho C_v}{1 + \chi \times \Delta t_3 / d_3^2}$$
$$= \frac{3.97 \times 10^6 \times 0.368 \times 10^9 \times 0.64 \times 10^{-6} / 2.77 \times 10^3 \times 0.88 \times 10^3}{1 + 85.9 \times 10^{-6} \times 0.64 \times 10^{-6} / \left(3.97 \times 10^{-6}\right)^2}$$
$$= 85^0 \text{ K}$$

and the Zener–Hollomon parameter equals

$$Z = \dot{\gamma} \exp \frac{Q}{R \cdot T} = 3.97 \times 10^6 \exp \frac{84 \times 10^3}{8.31434 \times 288} = 6.82 \times 10^{21}$$

These calculations show that the values of the deformation and the temperature within the shear band turn out to be insufficient for initiation of the dynamic recrystallisation. However, the microstructural picture shows the presence of dynamic recrystallisation with a mean grain size of 3–5 μm. The possible mechanism for such a process of recrystallisation is discussed below.

12.2.2 38CrNi3MoV

Another series of single shock loading and shock reloading tests were performed for the 38CrNi3MoV complex alloyed steel. The time-resolved free surface velocity profiles and velocity variance are presented in Fig. 12.4. The duration of the non-uniform deformation where the mesoscale velocity variance is nonzero equals 180 ns. For the shock reloading conditions, the mean value of the velocity variance equals ~ 18 m/s, whereas the duration of non-uniform straining equals 400 ns (see Fig. 12.4b):

The physical characteristics of steel are provided in Table 12.3. The shear band width, deformation, deformation rate and temperature were calculated with the same method as for D16 aluminium alloy. The temperature conductivity coefficient for 38CrNi3MoV steel equals

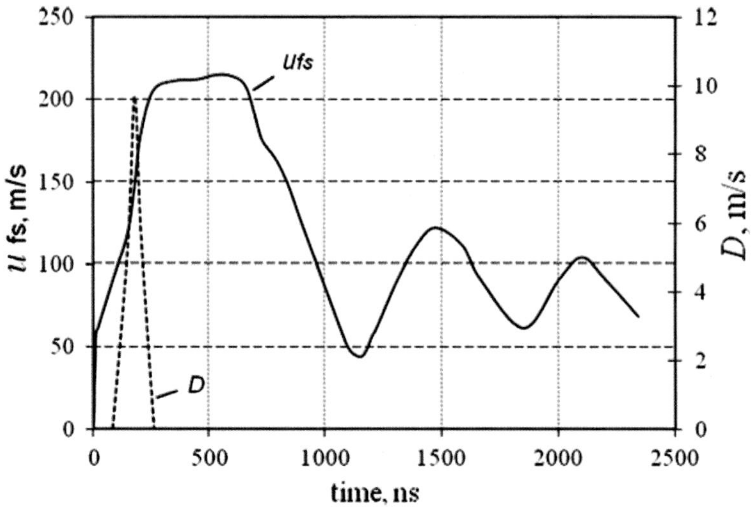

Fig. 12.4 Time-resolved free surface velocity profile, $u_{fs}(t)$, and velocity variance profile, $D(t)$, measured in tests under single loading shock conditions for 38CrNi3MoV steel

$$\chi = \frac{\lambda}{C_v \rho} = \frac{45.4}{0.448 \times 10^3 \times 7850} = 12.9 \times 10^{-6}\text{m}^2/\text{c}$$

In both cases, the width of the shear band equals

$$d_1 = \sqrt{\chi \frac{\Delta t_2}{2}} = \sqrt{\frac{12.9 \times 10^{-6} \times 0.18 \times 10^{-6}}{2}} = 1.08 \times 10^{-6}\text{m}$$

$$d_2 = \sqrt{\chi \frac{\Delta t_1}{2}} = \sqrt{\frac{12.9 \times 10^{-6} \times 0.4 \times 10^{-6}}{2}} = 1.6 \times 10^{-6}\text{m}.$$

The rates of shear strain are

$$\dot{\gamma}_1 = \frac{D_1}{d_1} = \frac{18}{1.6 \times 10^{-6}} = 11.25 \times 10^6 \text{ s}^{-1} \quad \text{and}$$

$$\dot{\gamma}_2 = \frac{D_2}{d_2} = \frac{9.7}{1.08 \times 10^{-6}} = 8.98 \times 10^6 \text{s}^{-1},$$

and the shear deformation is

$$\gamma_1 = \dot{\gamma}_1 \cdot \Delta t_1 = 11.25 \times 10^6 \times 0.4 \times 10^{-6} = 4.5 \quad \text{and}$$

$$\gamma_2 = \dot{\gamma}_2 \cdot \Delta t_2 = 8.98 \times 10^6 \times 0.18 \times 10^{-6} = 1.62.$$

The local shear stresses are

$$\tau_1 = \rho \times \mathrm{C_l} \times \mathrm{D_2} = 7850 \times 6100 \times 9.7 = 0.46 \times 10^9\mathrm{Pa}, \quad \tau_2 = \rho \times \mathrm{C_l} \times \mathrm{D_l} = 7850 \times 6100 \times 18 = 0.86 \times 10^9\mathrm{Pa}.$$

The temperature inside the shear band equals

$$\begin{aligned}\Delta T_1 &= \frac{\gamma_2 \cdot \tau_2 \cdot \Delta t_2/\rho C_v}{1 + \chi \cdot \Delta t_2/d_2^2} \\ &= \frac{8.98 \times 10^6 \times 0.46 \times 10^9 \times 0.18 \times 10^{-6}/7.85 \times 10^3 \times 0.448 \times 10^3}{1 + 12.9 \times 10^{-6} \cdot 0.18 \times 10^{-6}/\left(1.08 \times 10^{-6}\right)^2} \\ &= 70.4\ \mathrm{K}\end{aligned}$$

and

$$\begin{aligned}\Delta T_2 &= \frac{\gamma_1 \cdot \tau_1 \cdot \Delta t_1/\rho C_v}{1 + \chi \cdot \Delta t_1/d_1^2} \\ &= \frac{11.25 \times 10^6 \times 0.86 \times 10^9 \cdot 0.4 \times 10^{-6}/7.85 \times 10^3 \times 0.448 \times 10^3}{1 + 12.9 \times 10^{-6} \times 0.4 \times 10^{-6}/\left(1.6 \times 10^{-6}\right)^2} \\ &= 364.9\ \mathrm{K}.\end{aligned}$$

The Zener–Hollomon parameter equals

$$Z = \dot{\gamma} \exp \frac{Q}{R \cdot T} = 11.25 \times 10^6 \exp \frac{174 \times 10^3}{8.31434 \times 288} = 4.09 \times 10^{38}$$

These calculations show that the deformation and the deformation rate inside the shear band for 38CrNi3MoV steel are sufficient for the dynamic recrystallisation to be initiated, but the temperature is too low. In spite of that fact, the microstructural investigations reveal the presence of dynamic recrystallisation regions in steel targets.

In Fig. 12.6, the grain size distribution for D16 aluminium alloy is presented.

The main conclusion which follows from the above experiments is that dynamic recrystallisation is absent under single shock loading and is present in the reloading regime. Thus, there is a correspondence between the grain size and the parameters of the velocity non-uniformity, such as the velocity variance and the time duration when the velocity variance is nonzero.

The above experiments show that dynamic recrystallisation is absent in single shock loading and present in reloading experiments. The main feature of shock-induced dynamic recrystallisation is the short duration. The well-known mechanisms of DRX cannot be applied to explain the above phenomenon, as they demand too long a time compared to the duration of the load stress pulse. In this situation, the following scenario for shock-induced dynamic recrystallisation can be suggested. We would like to shine a light on the recently discovered phenomenon of collective nanoshears [12]. It was shown in Chap. 4 that, under high strain rates, the spatial polarisation

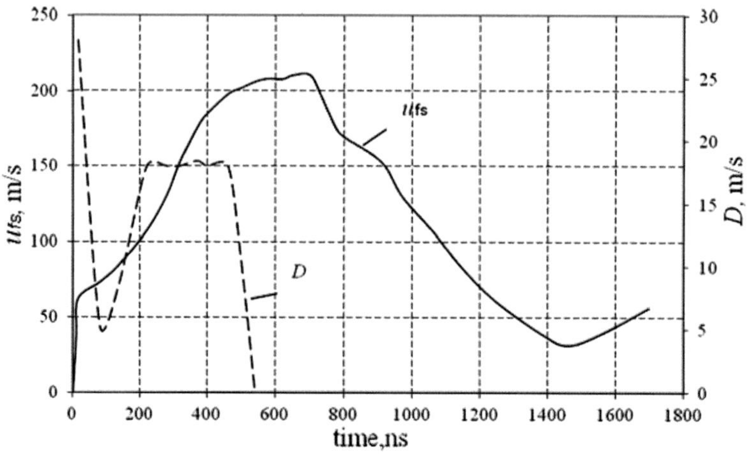

Fig. 12.5 Free surface velocity profiles, $u_{fs}(t)$, and velocity variance, $D(t)$, for 38CrNi3MoV steel target in a reloading regime

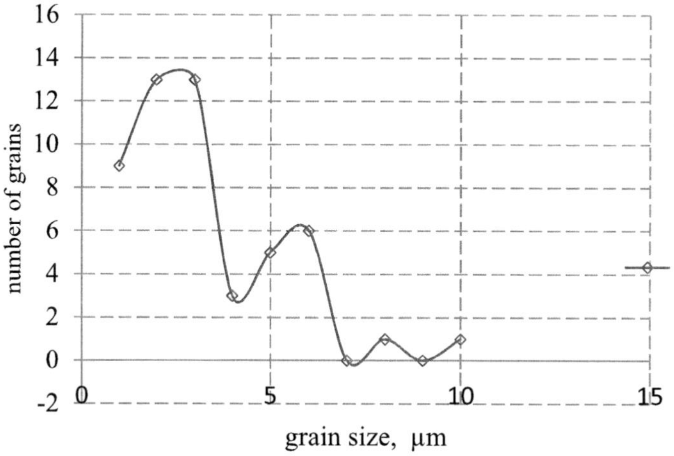

Fig. 12.6 Grain size distribution for D16 aluminium alloy

of the dislocation structure occurs: initially, chaotically distributed dislocations are rebuilt into quasi-periodical structures consisting of single-sign dislocation pile-ups. If the dislocation velocity is equal to the velocity of the transverse elastic waves in aluminium, $V_d = 3.2$ mm/μs, the spatial period of such a structure is equal to

$$L = V_d t_d = 3.2\,10^5 10^{-9} = 3.2\ \mu\text{m},$$

where $t_d = 10^{-9}$ s is the transient time for the return of the dislocation to a nominal velocity.

The next step is the formation of the dislocation cell structure. In the case of dynamic deformation, according to dislocation dynamics, the control parameter in rebuilding the polarised dislocation structure and creating the cell structure is the mesoscale velocity variance. The shear stress which initiates the rebuilding equals:

$$\sigma_{\text{II}} = 0.5\,\rho C_{pl} D_{\text{II}},$$

where $C_{\text{pl}}$is the velocity of plastic wave (for D16 aluminium alloy, $C_{\text{pl}}=$ 5.35 mm/μs) and D_{II} is the longitudinal component of the velocity variance. Excepting the value of velocity variance to 30 m/s (see Fig. 12.2), one obtains $\sigma_{\text{II}} =$ 217 MPa. The longitudinal component of stress, σ_{II}, initiates the formation of cell structure in the transverse slip planes of the crystal. A molecular dynamics simulation shows that the phase of the transverse component of the velocity variance is delayed relative to the longitudinal component by 90° [19] and initiates the generation of transverse local shear stress. This stress cannot influence the dislocations in primary slip systems. The local stresses in the secondary slip system only initiate the collective nanoshears, resulting in a quick rebuilding of the dislocation structure [12]. Each nanoshear is realised on distances of the order of ten atomic intervals, providing the reorientation of the lattice in the neighbouring regions of the crystal. The successive picture of such rebuilding is presented in Fig. 12.7. The generation of collective nanoshears is the third stage of the formation of shock-induced DRX flowing on very short timescales compared with the duration of the plastic front of the shock wave. The process of nucleation of the 3D structure is presented as a successive stage of rebuilding the initially chaotic dislocation structure into a polarised structure and resonance mixing.

12.3 Regimes of Shock Wave Propagation and Dynamic Recrystallisation

In this section, the coupling between the parameters of the shock wave and the initiation of dynamic recrystallisation is considered on the basis of experimental tests of D16 aluminium alloy. Shock loading of D16 aluminium alloy under uniaxial strain conditions was conducted in accordance with the method described in Chap. 8. The results of shock tests of D16 aluminium alloy are provided in Table 1, where the following designations are used: $h_{\text{imp}}$is the impactor thickness, U_{thr} is the free surface velocity corresponding to the change of inclination of the plastic front, U_{fsmax} is the maximum free surface velocity, δu_{def} is the velocity defect at the top of the compressive pulse, and W is the pull-back velocity. In Fig. 12.8, the dependencies for $W = f(U_{\text{imp}})$, $\delta u_{\text{def}} = f(U_{\text{imp}})$ and $D = f(U_{\text{imp}})$ are plotted. The increase in the velocity defect δu_{def} begins at impact velocities over 382 m/s. Note that the velocity variance becomes equal to the velocity defect ($\delta u_{\text{def}} = D$) at two values of impact

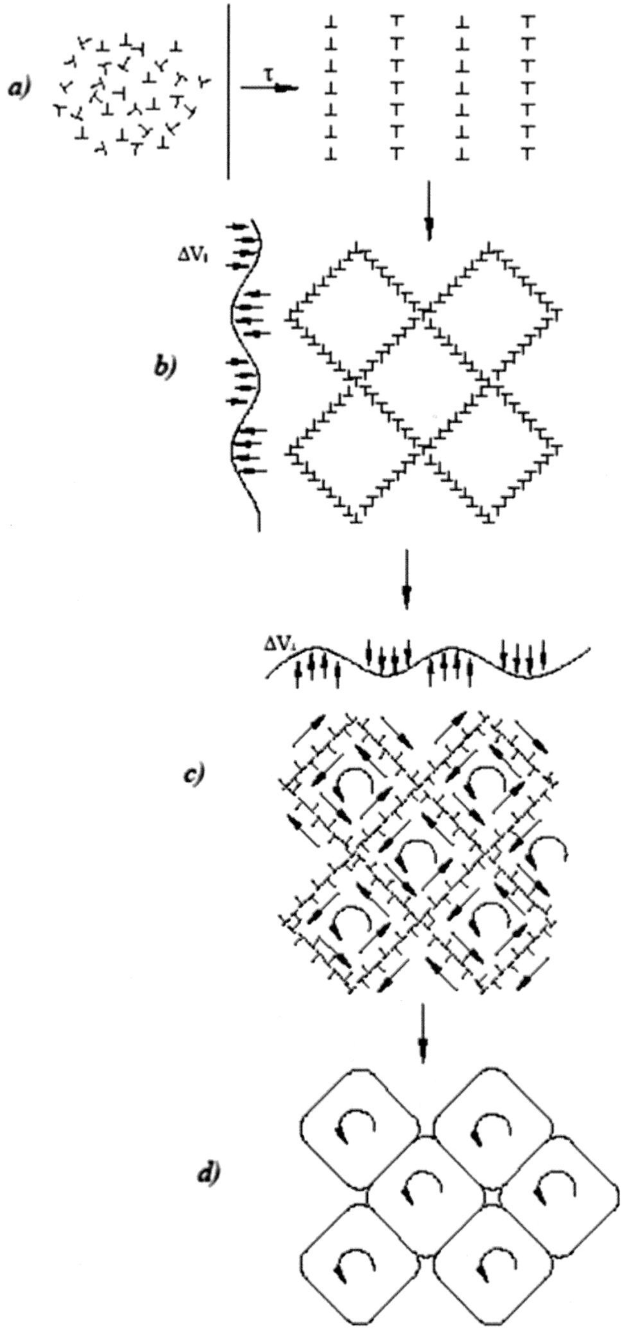

Fig. 12.7 Successive images of the formation of **a** the periodic polarised dislocation structure, **b** the cell structure, **c–d** the recrystallised structure

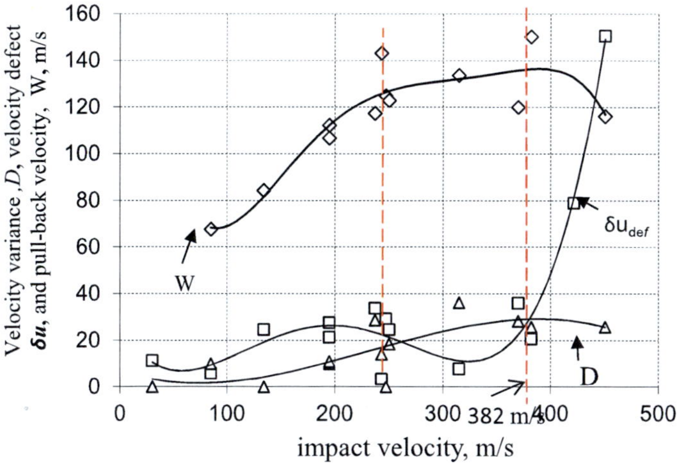

Fig. 12.8 Dependencies of the pull-back velocity, *W*, the velocity variance, *D*, and the velocity defect, δu_{def}

velocity, $U_{\mathrm{imp}} = 265$ m/s and $U_{\mathrm{imp}} = 382$ m/s. Both cases correspond to an increased value of the pull-back velocity.

In order to identify the mechanisms of structural transition and change in the regime of dynamic deformation, the microstructural and X-ray diffraction investigations of D16 aluminium alloy were performed. An estimate of the structural state of the metal was conducted on cross sections of targets parallel to the shock wave propagation. During these X-ray investigations, the intensities of the diffraction lines were fixed, which allows the identification of two orders of reflections from the (111) and (100) planes. The first family is the response of the planes of the lightest sliding, whereas the second family is an accommodation response.

The initial structure of the material is typical for a D16 aluminium alloy after cold deformation and thermal treatment. It is presented by 15–35 μm grains filled with small (~5 μm) particles surrounded by large (up to 20 μm) inclusions of Al_2CuMg phase. At an impact velocity of 265 m/s, the structural state outside the spall zone does not differ very much from the initial state. The structural state of the material changes more substantially at an impact velocity of 382 m/s. For comparison, in Fig. 12.9, the micrographs of the cross sections of the target loaded at the impact velocities of 265 m/s (Fig. 12.9a) and 382 m/s (Fig. 12.9b) are presented.

The cross section of the target loaded at 382 m/s is filled with 3D structures having longitudinal and transverse deformation lines (Fig. 12.9b). The boll formations of mesoscopic scale appear near the intermetallide inclusions. After etching and coating of the cross section by oxidising, the deformation structures are visualised. When the impact velocity increases to 422 m/s and higher, up to 451 m/s, the deformation structures disappear. Thus, the resonance character of the nucleation of 3D structures at an impact velocity of 382 m/s is available. Three-dimensional structures are nucleated for a very short time, on the order of a hundred nanoseconds.

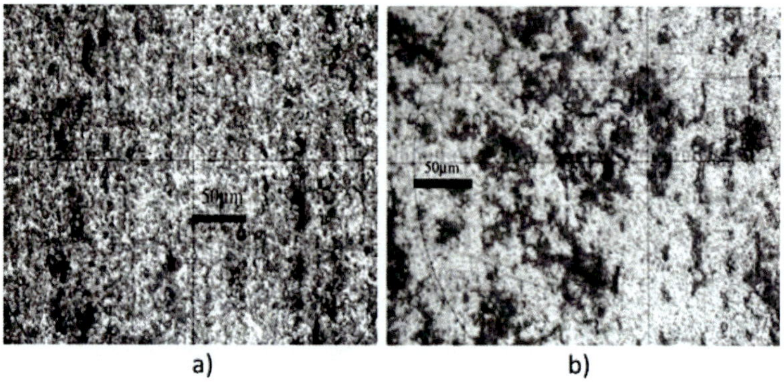

Fig. 12.9 Micrographs of D16 Al alloy structure after shock loading at an impact velocity of **a** 256 m/s and **b** 382 m/s

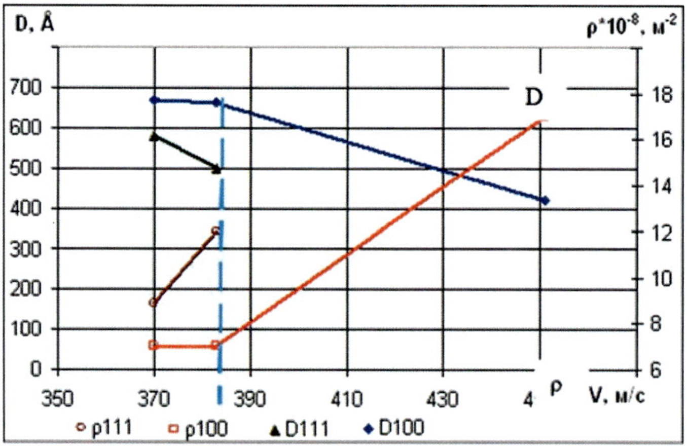

Fig. 12.10 Dependencies of the density of dislocations, ρ, and the mean size of regions of coherent scattering, *D,* on the impact velocity at planes <111> and <100)

X-ray diffraction analysis and microstructural investigations support a preconcerted behaviour for two structural characteristics—the density of dislocations and the regions of coherent scattering—depending on the crystallographic systems and the range of impact velocities. The density of dislocations in the (111) planes monotonously increases with increased impact velocity (see Fig. 1.10). Growth of dislocation density in (100) planes is also found, but this process flows at the impact velocities higher than 382 m/s. An analogous "threshold" trend was found for regions of coherent scattering. The curves indicate decrease from 650 to 400 Å at impact velocities over 382 m/s, reflecting the decreasing grain size (see dotted line in Fig. 12.10).

The cross section of the target loaded at 382 m/s is filled with 3D deformation structures, which have longitudinal and transverse components of deformation (Fig. 12.9b). The etching and coating of the cross section by oxidising provide a visualisation of the deformation structures. When the impact velocity increases to 422 m/s and over, up to 451 m/s, the deformation structures disappear. It may be concluded that the nucleation of the 3D structures flows under resonance conditions at an impact velocity of 382 m/s. The main feature of dynamic 3D structures is that they are nucleated for a very short time, on the order of a hundred nanoseconds. In Fig. 12.13, the process of nucleation of 3D structures is presented as successive stages of rebuilding the initially chaotic dislocation structure into a polarised structure and resonance mixing.

Considering the above results, we focus on collective nanoshears initiated by dynamic deformation [12]. The mechanism of the nucleation of 3D structures is explained using the concept of collective rebuilding of dislocation structures occurring under the resonance conditions of dynamic deformation. The controlling parameter in the rearrangement of the structure is the particle velocity variance, D. The driving stress responsible for the rearrangement equals

$$\sigma = 0.5\,\rho C_{\mathrm{pl}} D,$$

where $C_{\mathrm{pl}}$is the velocity of the plastic wave and D_{ms1} is the particle velocity variance at mesoscale 1. For D16 aluminium alloy, C_{pl}= 5.35 mm/μs. Taking the velocity variance as 30 m/s (see Fig. 12.8), one obtains $\sigma = 217$ MPa.

The dynamic recrystallisation flows for a very short time, which is comparable to the duration of the plastic front of the compressive pulse. This process flows in the form of collective nanoshears providing a discrete disorientation of sub-grains relative to each other.

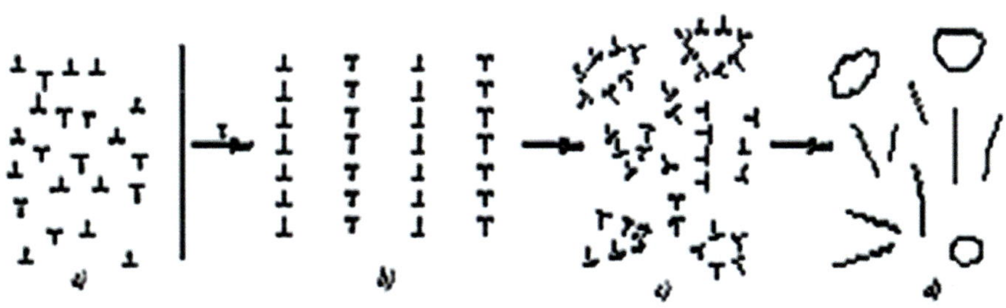

Fig. 12.11 Schematic of the nucleation of nanoshears under dynamic loading

References

1. Humphreys FJ, Hatherly M (1995) Recrystallization and related annealing phenomena. Pergamon Press, Oxford
2. Meyers MA, Nesterenko VF, La Salvia JC, Xue Q (2001) Mater Sci Eng A 317:204
3. Meyers MA, Xu YB, Xue Q, Pe'rez-Prado MT, Mc Nelley TR (2003) Acta Mater 51:1307
4. Meyers MA, Pak H-R (1986) Acta Met 34:2493
5. Hynes J, Vecchio KS, Ahzi S (1998) Met Mat Trans A 29:191
6. Liu Q, Hansen N (1995) Scripta Metall Mater 32:1289
7. Hughes DA, Chrzan DC, Liu Q, Hansen N (1998) Phys Rev Lett 81:4664
8. Hughes DA, Hansen N (1997) Acta Mater 45:3871
9. Sevillano JG, van Houtte P, Aernoudt E (1981) Prog Mat Sci 25:69
10. Meyers MA, LaSalvia JC, Nesterenko VF, Chen YJ, Kad BK (1944) Recrystallization and related phenomena. J H J Appl Phys 15:22–32
11. Gao W, Belyakov A, Miura H, Sakai T (1999) Mater Sci Eng A 265:213
12. Miroyama M, Howe JM, Hidaka JM, Takaki S (2003) ISIJ Int 43:755–760

Chapter 13
Multiscale Mechanisms of Dynamic Deformation Under High-Velocity Penetration

Abstract 1565 aluminum alloy has been tested in two schemes of shock loading: (i) shock loading under uniaxial strain conditions (plane collision of impactor and target), and (ii) high-velocity penetration of elongated rods. It is found that transition of dynamic deformation from mesoscale- (1-10 μm) on to mesoscale-2 (50-500 μm) in both schemes of loading happens at the identical impact velocities. The formation of large-scale structures of mesoscale-2 in form of fault structures at the lateral surfaces of penetration cavern increases the resistance of material to high velocity penetration and decreases the penetration length of elongated rod. At the same time, the transition on to large structural scale (mesoscale-2) decreases the spall strength of material.

13.1 Introduction

In the present chapter, the mesostructure formation is studied with combined experiments. A critical step in having an efficient picture of multiscale processes is parallel tests using two schemes of shock loading. The first scheme is under uniaxial strain conditions and the second scheme is the high-velocity penetration of an elongated rigid rod. In the second scheme, the impactor faces two mechanisms of resistance: the forehead resistance of the plane nose of the impactor and the friction resistance of the lateral region of the rod. In shock tests under uniaxial strain conditions, the target suffers only forehead resistance. Comparison of the results in both schemes reveals the role of macro–mesomomentum exchange in multiscale deformation and fracture.

Shock tests under uniaxial strain conditions were conducted using the experimental technique described in Chaps. 8 and 9. All the shock wave experiments were conducted with the 1565 aluminium alloy, the mechanical characteristics of which are provided in Table 13.1. Data on the dynamic strength of the material, including the dynamic yield limit, the spall strength and the threshold of structural instability, are inferred from the temporal profiles of the free surface velocity, $u_{fs}(t)$, which are registered with the velocity interferometer. This interference technique allows both

Y. Meshcheryakov, *Multiscale Mechanics of Shock Wave Processes*, Shock Wave and High Pressure Phenomena, https://doi.org/10.1007/978-981-16-4530-3_13

the mean particle velocity, u_{fs}, and the particle velocity variance, D, to be registered in real time (see Chap. 9).

The high-velocity penetration tests were conducted at the same facility. To ensure the impactor is perpendicular relative to the plane target, the 20 mm long rod of 5 mm diameter is mounted into a polyvinyl carbonate sabot. The conditions for "rigid rod and target" are provided by using high-strength 02Cr18Co9Mo5-VI maraging steel for the rod. The tests on penetration allow the evolution of structure depending on the impact velocity to be retraced owing to post-shock structural investigations of the targets loaded under different impact velocities. All of the post-shock targets were investigated with an Axio-Observer $Z-1$ m microscope after polishing and etching.

13.2 Structural Instability and Spall Strength

In the first series of shock tests under uniaxial strain conditions, the dependence $U_{fsmax} = f(U_{imp})$ over an impact velocity range of 250–750 m/s was studied. The critical changes of the slope of the curve happen at impact velocities of 440 and 625.5 m/s. In Fig. 13.1, the point of change of the slope is indicated by dotted lines.

In Fig. 13.2, the dependencies of the maximum free surface velocity, U_{fsmax}, and the spall strength, W, on the impact velocity are plotted together. For both curves, the breaks happen at the same impact velocities, which are indicated with dotted lines.

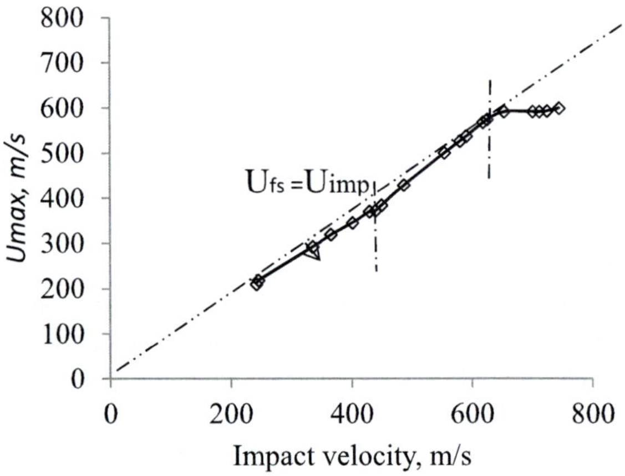

Fig. 13.1 Maximum free surface velocity U_{fsmax} versus impact velocity

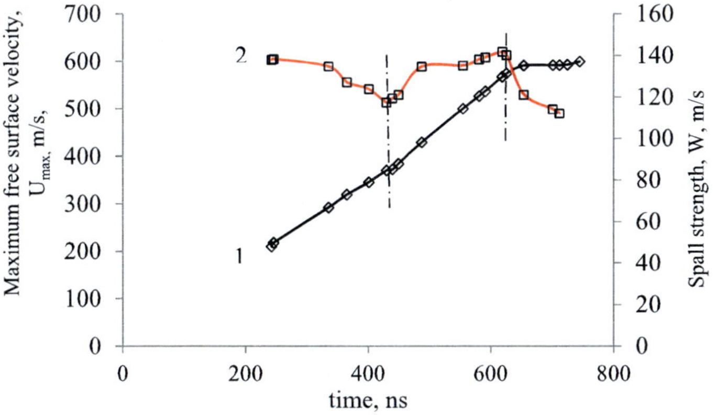

Fig. 13.2 Maximum free surface velocity, U_{fsmax} (1), and spall strength, W (2), versus impact velocity

13.3 The Structural Instability Threshold and High-Velocity Penetration

The resistance to high-velocity penetration is characterised, firstly, by the value of penetration depth, L, and secondly, by the slope of the curve $L = f(U_{\mathrm{imp}})$, namely (i) the smaller the penetration depth L, the higher the resistance to penetration and (ii) the smaller the slope of the curve $L = f(U_{\mathrm{imp}})$, the higher the resistance to penetration.

In Fig. 13.3, the dependencies of $U_{\mathrm{fsmax}} = f(U_{\mathrm{imp}})$ and the penetration depth $L = f(U_{\mathrm{imp}})$ on the impact velocity are plotted together. Within the impact velocity range of 241.9–750 m/s, the penetration curve experiences breaks at impact velocities of 440 m/s (position B') and 625.5 m/s (position C'). The relation between the rate of

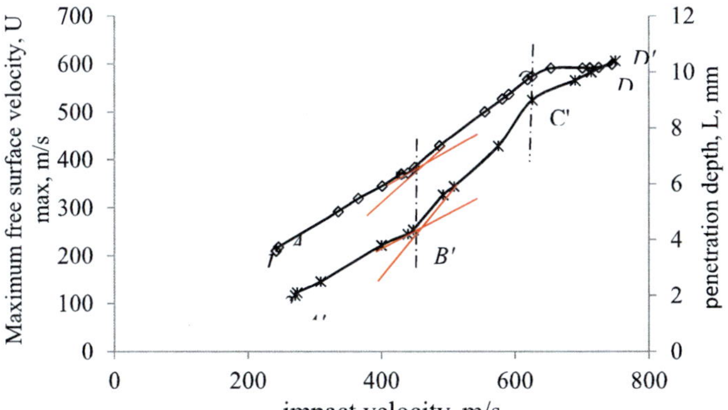

Fig. 13.3 Maximum free surface velocity, U_{fsmax} (1), and penetration depth, L (2), versus impact velocity

change of the structural instability threshold and the rate of change of the penetration depth can be summarised as follows:

1. The impact velocity range of 241.9–625.5 m/s: $\frac{dU_{max}^{AB}}{du} < \frac{dU_{max}^{BC}}{du}$; $\frac{dL_{A'B'}}{du} < \frac{dL_{B'C'}}{du}$.
2. The impact velocity range of 440–750 m/s: $\frac{dU_{max}^{CD}}{du} < \frac{dU_{max}^{BC}}{du}$; $\frac{dL_{C'D'}}{du} < \frac{dL_{B'C'}}{du}$.

The inequality $\frac{dL_{A'B'}}{du} < \frac{dL_{B'C'}}{du}$ means that the resistance to penetration decreases at impact velocities over 440 m/s. Within the impact velocity range from 625.5 to 750 m/s, after point C', an analogous inequality occurs: $\frac{dL_{C'D'}}{du} < \frac{dL_{B'C'}}{du}$, which means that the resistance to penetration at impact velocities over 625.5 m/s increases. It may be concluded that the processes responsible for the structural instability of the material and the high-velocity penetration are related.

13.4 Resistance to High-Velocity Penetration and Velocity Defect

Figure 13.4 shows that the behaviour of the velocity defect correlates with the behaviour of the penetration depth. The relations between the rate of change of the velocity defect and the rate of change of the penetration depth can be summarised as the following:

1. The impact velocity range 241.9–625.5 m/s: $\frac{d\delta u_{A'B'}^{def}}{du} > \frac{d\delta u_{B'C'}^{def}}{du}$, $\frac{dL_{AB}}{du} < \frac{dL_{BC}}{du}$.
2. The impact velocity range of 440–750 m/s: $\frac{d\delta u_{C'D'}^{def}}{du} > \frac{d\delta u_{B'C'}^{def}}{du}$. $\frac{dL_{CD}}{du} < \frac{dL_{BC}}{du}$.

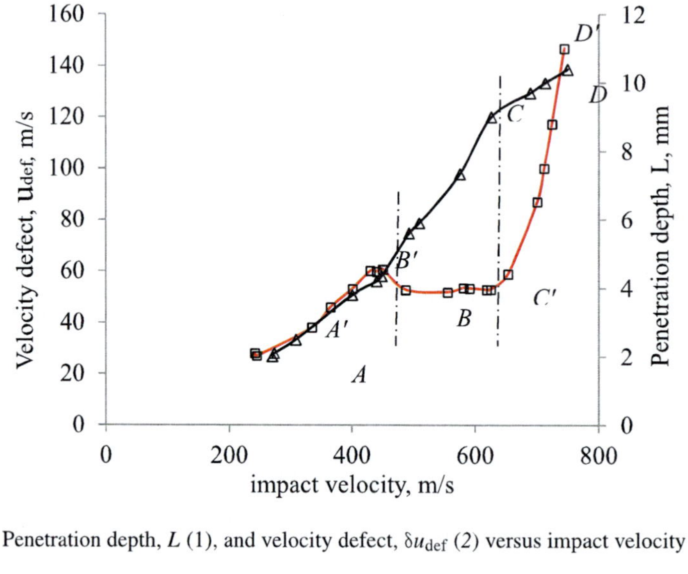

Fig. 13.4 Penetration depth, L (1), and velocity defect, δu_{def} (2) versus impact velocity

Within the upper range of impact velocities, *CD*, the resistance to penetration increases. The breaks for both curves happen at the same impact velocities (indicated with vertical dotted lines in Fig. 13.4).

13.5 Spall Strength, Resistance to High-Velocity Penetration and Velocity Variance

It is very interesting to compare the spall strength velocity variance and the resistance to penetration for different impact velocities. In Fig. 13.5, the penetration depth curve $L = f(U_{\text{imp}})$ is plotted with the pull-back velocity $W = f(U_{\text{imp}})$. Comparison of these curves shows that a correlation between these processes exists. The dependence $W = f(U_{\text{imp}})$ determined in the tests under uniaxial strain conditions experiences two breaks at the impact velocities of 440 and 625.5 m/s. The dependence of the penetration depth $L = f(U_{\text{imp}})$ also experiences two breaks at the impact velocities of 440 and 608 m/s. The critical change of slope for the penetration dependence happens at the impact velocity where the break for dependence $W = f(U_{\text{imp}})$ occurs (dotted lines in Fig. 13.5). Such behaviour of curves shows the common mechanism of structural behaviour in tests under uniaxial strain conditions and high-velocity penetration. The high impact velocity breaks in both loading schemes also happen at close values: 625.5 m/s in plane tests and 608 m/s in penetration tests. This means that both breaks are of the same nature.

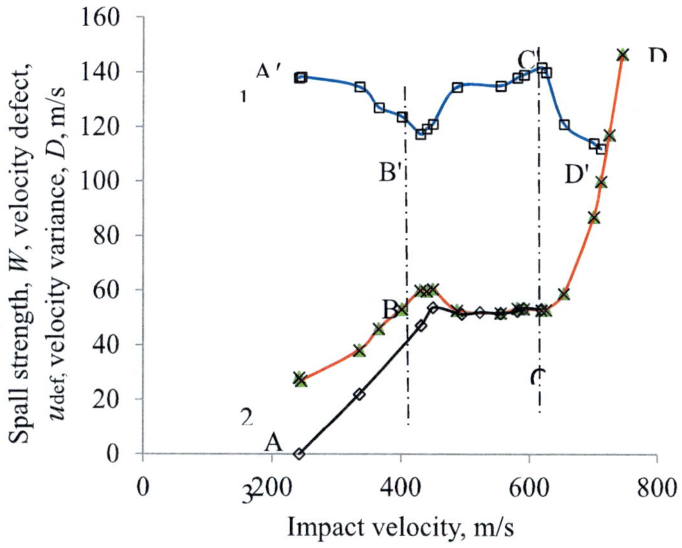

Fig. 13.5 Spall strength, W, (1), velocity defect, δu_{def}, (2) and velocity variance, D, (3) versus impact velocity

Now let us consider a correlation between the spall strength behaviour and the resistance to penetration over impact velocity ranges:

1. *Penetration*:
(a) Impact velocity range 241.9–625.5 m/s: $\frac{dL_{AB}}{du} < \frac{dL_{BC}}{du}$.
(b) Impact velocity range 440–750 m/s: $\frac{dL_{CD}}{du} < \frac{dL_{BC}}{du}$.

2. *Spallation.*
(a) Impact velocity range 241.9– 440 m/s: $\frac{dW_{A'B'}}{du} < 0$.
(b) Impact velocity range 440–625.5 m/s: $\frac{dW_{A'B'}}{du} \cong 0$.
(c) Impact velocity range 625.5–750 m/s: $\frac{dW_{A'B'}}{du} < 0$.

Within the impact velocity range of 241.9–625.5 m/s, the slope of the penetration depth curve changes at point B': $\frac{dL_{A'B'}}{du} < \frac{dL_{B'C'}}{du}$, which means that the resistance to penetration within the impact region $A'B'$ of curve (2) is higher compared to that for part $B'C'$. At the same time, the spall strength within the same impact velocity range shows the opposite trend. Within part AB, the spall strength decreases from 137.8 to 117.3 m/s. After point B, within part BC, $\frac{dW_{BC}}{du} \simeq 0$, which means that the spall strength is approximately constant. The critical changes in both curves happen at an impact velocity of 440 m/s. A similar situation is seen after the second critical impact velocity of ~625.5 m/s: within part C′D′ the slope of the penetration curve (2) decreases: $\frac{dL_{C'D'}}{du} < \frac{dL_{B'C'}}{du}$; that is, the resistance to penetration increases within region $C'D'$, whereas $\frac{dW_{CD}}{du} < 0$, which means that the spall strength in this region of strain rates decreases.

Thus, within the impact velocity range of approximately 241.9–750 m/s, the strength behaviour of 1565 aluminium alloy in two schemes of shock loading turns out to be opposite: when resistance to penetration increases, the spall strength decreases.

In Fig. 13.5, in addition to the spall strength and penetration depth, the dependence of the velocity defect on the impact velocity is provided. In our experiments, the mean particle velocity, u, and velocity variance, D, are registered in the form of temporal profiles. Independent measurements of the impact velocity allow three shock wave characteristics to be determined in real time. The breaks at curves are seen to occur at identical impact velocities. The spall strength is seen to decrease when the velocity defect increases (in the regions $A'B'$ and $C'D'$). At the same time, the spall strength is practically constant when the velocity defect remains constant.

Such spall strength behaviour has been analysed in Chap. 11, from the perspective of macro–mesomomentum exchange. In the case of equilibrium, when the rate of change of the velocity variance equals the rate of change of the mean particle velocity,

$$\frac{dD}{dt} = \frac{du}{dt} \tag{13.1}$$

the velocity defect equals the velocity variance:

$$\delta u_{\text{def}} = D \tag{13.2}$$

In the case, when the local strain rate equals the macroscopic strain rate, the dynamic deformation corresponds to the equilibrium regime. These conditions are accomplished in the middle range of impact velocities, B′C′ in Fig. 13.5. In shock wave experiments, the local strain rate cannot be determined, whereas the velocity defect and velocity variance can be determined. This allows the optimum regimes of dynamic deformation to be easily determined in experiments under uniaxial strain conditions.

13.6 Microstructural Investigations

The initial structure of 1565 aluminium alloy contains elongated grains, or texture (Fig. 13.6a). To understand the multiscale mechanisms of the strength behaviour of the alloy in different ranges of impact velocities, it is appropriate to compare the microstructural data for different ranges of impact velocities with the dependencies of penetration length over the same ranges. In accordance with Fig. 13.5, the overall range of impact velocities can be divided into three sub-zones where the strength behaviour of the material is different. The structural investigations show that the microstructure is also different for each range of impact velocities.

The main feature of post-shock specimens is the presence of numerous microshears of 3–10 μm in length oriented along the shock direction. The investigations of multiscale mechanisms of dynamic deformation in Chaps. 3 and 4 show that the behaviour of the mesostructure under dynamic loading can be characterised by the presence of short-lived pulsations of particle velocity. For example, the motions of

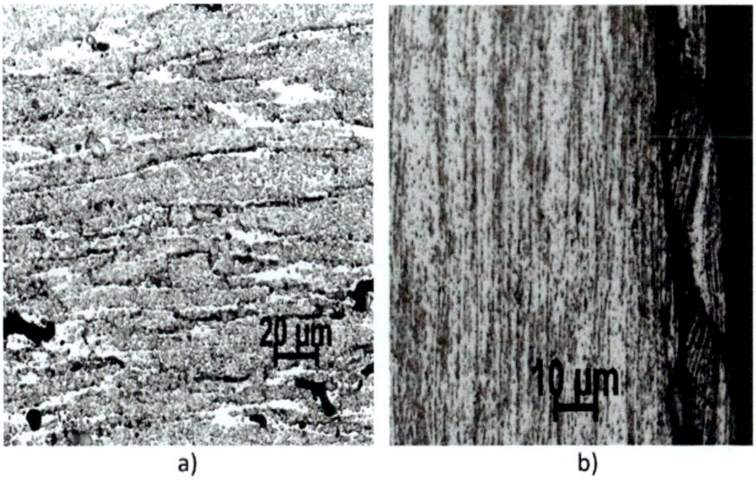

Fig. 13.6 Initial structural states of **a** 1565 aluminium alloy, and **b** microshears and fault structures in a post-shock specimen

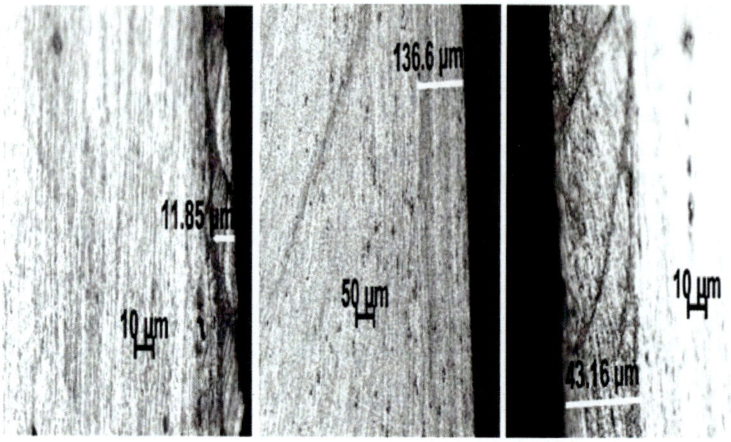

a) Region AB b) Region BC c) Region CD

Fig. 13.7 Three regions of lateral structures in a 1565 aluminium alloy target after penetration. The microstructure of region *AB*

the mesostructural elements in the form of single-sign dislocation groups in D16 aluminium alloy create the short-livid velocity pulsations of $\Delta t \approx 150$–200 ns. The velocity interferometer registers the particle velocity distribution in the form of velocity variance *D*. In the case of 1565 aluminium alloy, the maximum value of velocity variance at mesoscale 1 equals $D = 4.5 \times 10^3$ cm/s, from which the mean displacement equals: $\mathrm{L} = \mathrm{D} \cdot \Delta t = 4.5 \times 10^3 \cdot 2 \times 10^{-7} = 9 \times 10^{-4}$ cm. These displacements belong to mesoscale 1 (1–10 μm). With decreasing the distance to bank of cavern, the density of microshears increases.

Another structural element in high-velocity penetration is the fault structures at the back of the cavity. In Fig. 13.6b, a complex morphology of fault structures is presented: inside each cell, a family of shear bands is clearly visible.

In Fig. 13.7, the structural states for different ranges of impact velocity are provided. Within the regions *AB* and *CD*, a specific feature of the post-shock specimens is the presence of regular fault structures near the back of the cavity and uniform deformation in the region *BC*. The width of the region occupied by fault structures depends on the impact velocity: within region*AB*, the width of the fault structure zone equals 7–12 μm, whereas in the region *CD*, the width of the zone increases to 47–50 μm. Comparison with the deformation regions for the high-velocity penetration curve (Fig. 13.5) shows that the resistance to penetration within regions *AB* and *CD* increases. Within region *BC*, where the regular faults are absent, the post-shock structure is uniform whilst the resistance to penetration decreases. From the point of view of resistance to penetration, the 1565 aluminium alloy turns out to be preferable at the upper range of impact velocities. At the same time, the spall strength of 1565 alloy in the upper range of impact velocities decreases (see Fig. 13.5).

It has been shown in Chap. 7 that the temporal resolution of the experimental diagnostics (0.6–1.0 ns) allows measurement of the fine structure of the plastic wave, including the high-frequency oscillations at mesoscale 2. Where observed, the oscillations are excited at the top of the plastic front. Figure 7.1 of Chap. 7 demonstrates the oscillations in a 7 mm 1565 aluminium alloy target loaded at an impact velocity of 636.5 m/s. The oscillations are seen only at the impact velocities higher than the threshold of the structurally unstable transition, U_{ins}. The spatial period of oscillations equals ~50 μm, which coincides with the mean size fault structures (40–50 μm). Thus, whilst the mesoscale 1 structures are nucleated owing to particle velocity pulsations, the fault structures of mesoscale 2 are initiated due to the resonance interaction of the mesoscale 2 structures with the plastic front oscillations, which, in turn, result from positive back coupling between the mesoparticle velocity dispersion (mesoscale) and the velocity defect (macroscale).

Below, the dimensions of fault structures are calculated for regions *AB* and *CD* using the relationships (7.22) and (7.24) of Chap. 7.

Region AB.

1. Impact velocity $U_{\text{imp}} = 335$ m/s
2. Velocity variance $D = 22$ m/s.
3. Velocity of plastic front $C_p = 5.478 \times 10^5$ cm/s.
4. Longitudinal sound velocity $C_l = 6.387 \times 10^5$ cm/s.
5. Strain rate $\frac{\text{d}\varepsilon}{\text{d}t} = 7.79 \times 10^5\ \text{c}^{-1}$.
6. Shear modulus $\mu = 2.7 \times 10^{11}\,\frac{\text{dyn}\cdot\text{cm}}{\text{cek}^2}$.
7. Density $\rho = 2.7$ g/cm^3.

From Eq. (7.22), $R = \frac{D}{\text{d}\epsilon\ /\text{d}t} = 28 \times 10^{-4}$ cm and from Eq. (7.24), $\omega = 2.35 \times 10^8\ \text{c}^{-1}$ or $f = \omega\ /2\pi = 0.374 \times 10^8\ \text{c}^{-1}$, which corresponds to a time period of oscillation $T = \frac{1}{f} = 2.672 \times 10^{-8}$ c. Then the spatial period of the cell structures in the region *AB* equals: $\lambda = T \times U_{\text{imp}} = 2.672 \times 10^{-8} c \times 3.35 \times 10^4 \text{cm}\ /\text{c} = 8.95 \times 10^{-4}$ cm. The obtained values for the spatial period of the oscillations coincide with the dimensions of the fault structures shown in Fig. 13.7a.

Region CD.

1. Impact velocity $U_{\text{imp}} = 653.7$ m/s.
2. Velocity variance $D = 52$ m/s.
3. Velocity of plastic front C_p=5.66×10^5 cm/s.
4. Longitudinal sound velocity C_1=6.387×10^5 cm/s.
5. Strain rate $\frac{\text{d}\epsilon}{\text{d}t} = 5 \times 10^{-6}\ \text{c}^{-1}$.
6. Shear modulus $\mu = 2.7 \times 10^{11}\,\frac{\text{dyn}\times\text{cm}}{\text{cek}^2}$.
7. Density $\rho = 2.7$ g/cm^3.

From Eq. (7.22), $R = \frac{D}{\text{d}\epsilon\ /\text{d}t} = 10.36 \times 10^{-4}$ cm $\omega = 1.84 \times 10^8 c^{-1}$ or $f = \omega\ /2\pi = 0.29 \times 10^8 c^{-1}$, which corresponds to the time period of oscillation $T = \frac{1}{f} = 2.35 \times 10^{-8}$ c. Then the dimension of the fault structures in region *CD* equals: $\lambda = T \times U_{\text{imp}} = 2.135 \times 10^{-8} \times 6.637 \times 10^4 \text{cm}\ /\text{c} = 47 \times 10^{-4}$ cm. The obtained value for the spatial period of oscillations coincides with the dimensions of the fault structures for region *CD* shown in Fig. 13.7c.

13.7 Conclusions

- Shock tests of 1565 aluminium alloy using two loading geometries reveal a correlation between threshold mechanisms and scales of mesostructure formation.
- The maximum dynamic strength of the material corresponds to the strain region where the velocity defect equals the velocity variance.
- The formation of fault structures at the back of the penetration cavity increases the resistance to high-velocity penetration.
- The strength behaviour of the alloy in these two loading schemes proved to be opposite—when the resistance to penetration increases, the spall strength decreases.
- The defect of particle velocity acts as a control parameter in the dynamic deformation of the material and defines the transition from one scale to another.